Lecture Notes in Physics
Monographs

Springer
Berlin
Heidelberg
New York
Barcelona
Hong Kong
London
Milan
Paris
Tokyo

Physics and Astronomy

ONLINE LIBRARY

http://www.springer.de/phys/

The Editorial Policy for Monographs

The series Lecture Notes in Physics reports new developments in physical research and teaching - quickly, informally, and at a high level. The type of material considered for publication in the monograph Series includes monographs presenting original research or new angles in a classical field. The timeliness of a manuscript is more important than its form, which may be preliminary or tentative. Manuscripts should be reasonably self-contained. They will often present not only results of the author(s) but also related work by other people and will provide sufficient motivation, examples, and applications.

The manuscripts or a detailed description thereof should be submitted either to one of the series editors or to the managing editor. The proposal is then carefully refereed. A final decision concerning publication can often only be made on the basis of the complete manuscript, but otherwise the editors will try to make a preliminary decision as definite as they can on the basis of the available information.

Manuscripts should be no less than 100 and preferably no more than 400 pages in length. Final manuscripts should be in English. They should include a table of contents and an informative introduction accessible also to readers not particularly familiar with the topic treated. Authors are free to use the material in other publications. However, if extensive use is made elsewhere, the publisher should be informed. Authors receive jointly 30 complimentary copies of their book. They are entitled to purchase further copies of their book at a reduced rate. No reprints of individual contributions can be supplied. No royalty is paid on Lecture Notes in Physics volumes. Commitment to publish is made by letter of interest rather than by signing a formal contract. Springer-Verlag secures the copyright for each volume.

The Production Process

The books are hardbound, and quality paper appropriate to the needs of the author(s) is used. Publication time is about ten weeks. More than twenty years of experience guarantee authors the best possible service. To reach the goal of rapid publication at a low price the technique of photographic reproduction from a camera-ready manuscript was chosen. This process shifts the main responsibility for the technical quality considerably from the publisher to the author. We therefore urge all authors to observe very carefully our guidelines for the preparation of camera-ready manuscripts, which we will supply on request. This applies especially to the quality of figures and halftones submitted for publication. Figures should be submitted as originals or glossy prints, as very often Xerox copies are not suitable for reproduction. For the same reason, any writing within figures should not be smaller than 2.5 mm. It might be useful to look at some of the volumes already published or, especially if some atypical text is planned, to write to the Physics Editorial Department of Springer-Verlag direct. This avoids mistakes and time-consuming correspondence during the production period.

As a special service, we offer free of charge LaTeX and TeX macro packages to format the text according to Springer-Verlag's quality requirements. We strongly recommend authors to make use of this offer, as the result will be a book of considerably improved technical quality.

For further information please contact Springer-Verlag, Physics Editorial Department II, Tiergartenstrasse 17, D-69121 Heidelberg, Germany.

Series homepage -- http://www.springer.de/phys/books/lnpm

Heike Emmerich

The Diffuse Interface Approach in Materials Science

Thermodynamic Concepts and Applications of Phase-Field Models

 Springer

Author

Heike Emmerich
Lehrstuhl für Thermodynamik
Fachbereich Chemietechnik
Universität Dortmund
Emil-Figge-Strasse 70
44227 Dortmund, Germany

Cover picture: **Columnar dendrites: The shading figure denotes the variation of temperature in the melt. Typically columnar dendrites grow at the container walls as well as at the macroscopic liquid-solid interface, whereas so called equiaxed crystals develop in the inner regions of the molten phase. (by H. Emmerich)**

Cataloging-in-Publication Data applied for

A catalog record for this book is available from the Library of Congress.

Bibliographic information published by Die Deutsche Bibliothek

Die Deutsche Bibliothek lists this publication in the Deutsche Nationalbibliografie;
detailed bibliographic data is available in the Internet at http://dnb.ddb.de

ISSN 0940-7677 (Lecture Notes in Physics. Monographs)
ISBN 3-540-00416-5 Springer-Verlag Berlin Heidelberg New York

Springer-Verlag Berlin Heidelberg New York
a member of BertelsmannSpringer Science+Business Media GmbH

http://www.springer.de

© Springer-Verlag Berlin Heidelberg 2003
Printed in Germany

Typesetting by the authors/editors
Camera-data conversion by Steingraeber Satztechnik GmbH Heidelberg
Cover design: *design & production*, Heidelberg

Printed on acid-free paper   55/3141/du - 5 4 3 2 1 0

Acknowledgements

At this point I would like to thank Prof. Dr. Schreiber for giving me the opportunity to teach at Chemnitz and to integrate my work in a university chair's. I am also thankful for his detailed comments on the manuscript helping me to improve its readability in several points.

Moreover I would like to thank Prof. Dr. Kassner for introducing me to the field of simulating growth phenomena in materials science. Prof. Dr. Müller-Krumbhaar, Dr. Brener, Dr. Misbah and Dr. Ihle gave valuable guidance, which I am thankful for.

I am also grateful to Prof. Dr. Fulde and Prof. Dr. Rost for the stimulation and the opportunities which arose from being a Distinguished Postdoctoral Fellow at the Max Planck Institute for the Physics of Complex Phenomena at Dresden.

Finally I would like to thank Prof. Dr. Sadowski for offering the quite challenging new task to continue in the field of materials science by bridging the gap from the mesoscale to the atomar scale rising a respective simulation group at Dortmund university.

Dortmund, November 2002   *Heike Emmerich*

Contents

1. **Introduction** 1
   1.1 Structure and Scope of This Work 4

2. **What Is an Interface?** 7

3. **Equilibrium Thermodynamics of Multiphase Systems:**
   Thermodynamic Potentials and Phase Diagrams 19
   3.1 Calculating Phase Diagrams
       from Energy Functionals 21
   3.2 Abstracted Phase Diagrams 23
       3.2.1 Constructing the Gibbs Free Energies 26
       3.2.2 Non-conserved Versus Conserved Variable Approach .. 28

4. **Thermodynamic Concepts of Phase-Field Modeling** 31
   4.1 Derivation of Transport Equations 32
       4.1.1 Considering Conserved Quantities Only 32
       4.1.2 Extension to Non-conserved Quantities 35
   4.2 Introducing the Phase-Field Variable Φ 38
   4.3 Thermodynamic Consistency 46
       4.3.1 The Gradient Flow Method 47
       4.3.2 Entropy Production in Terms of Transport Variables .. 49
   4.4 Appendix 53

5. **Asymptotic Analysis** 59
   5.1 A Formal Mathematical Approach Towards Matching 59
       5.1.1 Illustration of Extended Domains 63
       5.1.2 Matching in the Context of Diffuse Interface Modeling 67
   5.2 Asymptotic Matching for Thin Film Epitaxial Growth 68
       5.2.1 Motivation 69
       5.2.2 Sharp Interface Formulation 70
       5.2.3 The Diffuse Interface Model Equations
             and Their Asymptotic Analysis 72
   5.3 A Generalized Approach Towards the Asymptotic Analysis
       of Diffuse Interface Models 78

          5.3.1   Role of the Governing Thermodynamic Potential   78

          5.3.2   The Expansion Procedure in Detail   81

    5.4   Discussion   92

    5.5   Appendix   93

6.   Application of Diffuse Interface Modeling
to Hydrodynamically Driven Growth   97

    6.1   Diffuse Interface Modeling
for Hydrodynamically Influenced Growth   99

    6.2   The Selection Problem of Dendritic Growth Revisited 104

    6.3   Comparison to Experimental Data 119

    6.4   Summary 127

    6.5   Appendix 129

7.   Application to Epitaxial Growth Involving Elasticity 131

    7.1   Elastic Driving Forces within the Diffuse Interface Approach . 132

    7.2   Comparing Simulations and Experiments 134

    7.3   Discussion 140

8.   Conclusions and Perspectives 141

Appendix 145

A.   Numerical Issues of Diffuse Interface Modeling 145

    A.1   Relationship Between Physical Variables
and Numerical Parameters 146

    A.2   Computational Universality of Phase Field Models 151

    A.3   Selected State-of-the-Art Numerical Approaches 154

          A.3.1   2D Adaptive Mesh Refinement Computation 155

          A.3.2   3D Simulations Including Fluid Flow 161

References 165

Index 175

1. Introduction

Many inhomogeneous systems involve domains of well-defined phases separated by a distinct interface. If they are driven out of equilibrium one phase will grow at the cost of the other. Examples are phase separation by spinodal decomposition or nucleation and subsequent growth of the nucleus in the nourishing phase [139]. Another example which has often been discussed as a paradigmatic problem is that of dendritic solidification [29, 64, 79, 199]. The phenomenological description of these phenomena involves the definition of a precisely located interfacial surface on which boundary conditions are imposed. One of those boundary conditions typically yields a normal velocity at which the interface is moving. This is the so-called *sharp interface* approach, adopted both in analytical and numerical studies for a variety of contexts involving a moving boundary. The origin of such a description is often transparent, being obtained by symmetry arguments and common sense. Nevertheless the properties of sharp interface models can be quite subtle as in the case for dendritic growth. This is strongly coupled to the question of how to view the interfacial surface. Already when introducing the notion of a surface quantity Gibbs implicitly entertained the idea of a diffuse interface [126]: any density of an extensive quantity (e.g., the mass density) between two coexisting phases varies smoothly from its value in one phase to its value in the other. The existence of a transition zone, though microscopically of atomic extent, underlies this definition of surface quantities as given by Gibbs. In phase transition phenomena, this notion has been employed in the spirit of Landau and Khalatnikov [271], who were the first to introduce an additional parameter to label the different phases in their theory on the absorption of liquid helium. Essentially diffuse interface modeling, as it appeared subsequently in the literature in the context of phase transition phenomena [54, 143], is connected to such an additional order parameter. Clearly such models have advanced numerical treatment as well as understanding of interfacial growth phenomena since.

Even though quite a young approach to tackle such problems, diffuse interface models have been employed by different groups in quite different spirits. One might even be tempted to say that a variety of philosophies accompanying diffuse interface modeling have already emerged. One way to view this method to model interfacial growth is to understand it as a numerical technique, which helps to overcome the necessity of solving for the precise location of the interfacial surface explicitly in each time step of a numerical simulation. This can be achieved by the introduction of one or several additional *phase-field* variables. They are the key element to the resulting *phase-field* modeling approach for studying systems out of equilibrium. In such an approach the phase-field variables are continuous fields which are functions of space **r** and time t. They are introduced to describe the different relevant phases. Typically these fields vary slowly in bulk regions and rapidly, on length scales of the order of the correlation length ξ, near interfaces. ξ is also a measure for the finite thickness of the interface. The free energy functional $\mathcal{A}$ determines the phase behavior. Together with the equations of motion this yields a complete description of the evolution of the system. In other contexts, such as critical dynamics [41, 139, 146], the fields are order parameters distinguishing the different phases. In a binary alloy, for example, the local concentration or sub-lattice concentration can be described by such fields. The ideas involved in this approach have a long history, referring back to van der Waals [244].

On this background the materials science community associates the use of continuum field models in particular with the work of Cahn and collaborators [7, 54, 55]. Within their contribution to the field, phase-field models are more but just a "trick" to overcome numerical difficulties. Rather they are rigorously derived based on the variational principles of irreversible thermodynamics as founded by Onsager [227]. Then ensuring *thermodynamic consistency* of the model equations can serve as a justification of a phase-field model. In this sense phase-field models can also be formulated for problems, for which sharp interface equations are not yet available. Consequently it might be their analysis which yields a formerly unknown sharp interface formulation and helps to clarify the physics in the interfacial region.

One has to contrast this procedure to a very established second way to validate a phase-field model. This second approach assumes, that a given sharp interface formulation of the growth problem is the correct description of the physics under consideration. On the basis of this assumption, a phase-field model can be justified by simply showing that it is asymptotic to the correct sharp interface description, i.e., that the latter arises as the *sharp interface limit* of the phase-field model when the interface width is taken to zero. Obviously this procedure works only for cases, in which a well established set of continuum equations describing the dynamics in the sharp interface formulation does exist. Moreover, employed in this way phase-field models do not seem to be much help to elucidate the physics of the interfacial region beyond what is captured within the sharp interface model equations.

However, the latter is only partially true and leads to a third philosophy appearing in the phase-field community lately. It is rooted in the understanding of the interfacial surface to be finite in the sense of Gibbs denoted above: If one assumes a phase-field model to be thermodynamically consistent and to describe a physical situation, for which an established sharp interface formulation exists, as well, then, certainly, in the *sharp interface limit* the phase-field model should correspond precisely to that sharp interface formulation. However, keeping in mind that the interface can be understood to be of finite width, not only the *sharp interface limit* of a phase-field model is a meaningful physical limit, but also the so-called *thin interface limit* introduced by Karma and Rappel [168, 169]. To clarify the difference between the *sharp interface limit* and this *thin interface limit* here I will consider the growth of a dendrite with tip radius R into an undercooled melt [128]. Under more general circumstances, R might be representative of a typical macroscopic length scale such as the container size. For dendritic solidification at large undercoolings the growth is rapid and the radius of curvature of the dendritic tip is relatively small. As a consequence effects of capillary action and kinetics on the local interfacial temperature can be significant. In this regime, sharp interface limits of the phase-field equations have been performed [49, 60, 115, 116, 121, 213], which assume that the dimensionless interfacial temperature u is of the order of the small parameter ξ/R. Contributions from capillary effects and kinetics can be regarded to be of the same order. In this limit one also considers ξ to be small compared to the capillary length l_c, which presents a stringent resolution requirement for a numerical computation that aspires to describe this limiting case. At low undercoolings, on the other hand, dendrites grow more slowly and have a larger radius of curvature, so that it is reasonable to model capillary effects and kinetics as small corrections. Karma and Rappel refer to the corresponding analysis as the *thin interface limit*. For this thin interface limit one assumes $\xi \ll R$ but allows $\xi \sim l_c$. Almgren [11] has described this analysis as *isothermal asymptotics*, since to leading-order in ξ/R the temperature is isothermal throughout the interfacial region with $u = O(\xi/R)$.

Now again one interest in employing such an isothermal asymptotics or thin interface limit can be understood to be of numerical origin: it can serve to legitimate a choice of model parameters which ensures *better*[1] numerical performance. On the other hand, isothermal asymptotics can also be used to obtain first order generalizations of the well known Gibbs-Thompson relation, which usually yields the temperature value locally at the interface. In turn such a generalization can facilitate subsequent stability analysis of the model.

Thus momentarily diffuse interface modeling is a field, in which numerical efforts as well as an intense focus on thermodynamic backgrounds and asymptotic behavior of the models drives the development to turn this approach into a more and more powerful technique. Within the present work I will elucidate this symbiotic framework in more detail with respect to specific growth phenomena. In particular I will show that it is this symbiosis which opens up new perspectives to gain further understanding about interfacial growth problems, if one extends the paradigmatic, purely diffusion limited

---

[1] The meaning of *better* as used in the above context will become elucidated further in the numerical appendix of this book.

dendritic growth problem step by step to additional physical mechanisms such as hydrodynamics or elasticity. One might wonder if in the end this approach can provide a framework to tackle the behavior of still more complicated systems, e.g. systems with an inherent multi-scale nature due to an internal structure, such as liquid crystals or polymer solutions, as well.

1.1 Structure and Scope of This Work

To elucidate the above concepts arising in the context of diffuse interface modeling I will first – within Chaps. 3 and 4 – present the thermodynamic background underlying that kind of modeling. It allows me to derive the respective model equations on the basis of variational principles.

I will then proceed to introduce the mathematical formulation which allows me to establish their correspondence to the sharp interface models, i.e. the method of *asymptotic analysis*.

Regardless of whether one intends to employ a diffuse interface model (a) merely as a computational trick or (b) to derive new information of the physics in the interfacial region, it is an important tool to deal with diffuse interface models: In case (a) the analysis might impose restrictions onto some of the numerical parameters of the model. Moreover it serves as validation of the model. In case (b) the method of *asymptotic analysis* is an instrument to obtain expressions, which directly yield the velocity and the equilibrium conditions of the interface. In this case ensuring thermodynamic consistency of the models serves as their validation. Usually this has to be guaranteed in addition to the variational principles and the asymptotic formalisms involved in the constitution of the models. Only recently a generalized asymptotic analysis which inherently implies thermodynamic consistency has been introduced by Elder *et al.* [102]. The way it was introduced it works for a restricted class of interfacial growth problems, only. In this way it is described in detail in Sect. 5.3.

In Chap. 6 and Chap. 7 the asymptotic analysis is taken up and employed to address very recent issues involved with interfacial growth in the case the latter is influenced not only by diffuse transport in the bulk phases but by hydrodynamic transport and elastic driving forces as well.

Finally, in Chap. 8 this text concludes with a summary of the trends and the perspective which open up in this young scientific field of diffuse interface modeling in materials science.

To close the circle of discussion an appendix to the overall text provides an overview on the numerical, implementation related issues involved.

The general **scope of this book** is to provide material, the study of which will enable the reader to apply the diffuse interface approach to the interfacial growth phenomena he is interested in. Along this line the second chapter can be understood as an introduction to the notion of interfacial growth and the success achieved in the past modeling it based on a sharp interface approach.

Such a sharp interface formulation is the frame of reference for the discussion of the diffuse interface approach throughout this text. Thus the purpose of Chap. 2 is to demonstrate, that – even though phenomenological in nature – for the sharp interface approach it is possible to establish a thorough relation between theory and experiment. In that sense it is regarded a 'solid' basis it is justified to build upon.

The following Chaps. 3–5 develop the thermodynamic background as well as the mathematical tools required if one intends to engage into diffuse interface modeling oneself. To keep them stringent, they remain deliberately formal. However, they start from a level that any student of physics, mathematics or an engineering science could follow. For that purpose Appendix A of Chap. 4 puts together some basics of the calculus of variations as required to follow the derivation of model equations in Chap. 4 and succeeding chapters. Moreover, Appendix B and Appendix C relate the Cahn–Hilliard equation respective the Allen–Cahn equation, as fundamental model equations summarizing the concept of diffuse interface modeling, to some concrete physical phenomena they allow us to describe. Within Chap. 5 the formal theory of matching is placed into the context of physics by elucidating its application for the example of thin film epitaxial growth (Sect. 5.2).

Chapters 6 and 7 are the ones, in which all the formalism discussed before is applied to two examples, namely

1. hydrodynamically influenced dendritic growth and
2. elastically influenced epitaxial growth.

These two applications serve as examples for phenomena, where the extension of sharp interface models to grasp all features of the growth process is no longer clear. In that sense they are examples, for which the achievement of diffuse interface modeling is obvious at first sight: It allows to model and investigate these phenomena, for which a sharp interface approach would fail. Moreover, in these cases it is the analysis of the diffuse interface models, which reveals new physical insight, which is not clear how to obtain otherwise. Chapters 6 and 7 are also meant to serve as guidance for the reader to construct a diffuse interface model for the application he might have in mind.

Within the concluding Chap. 8 the discussion of achievements of diffuse interface modeling as examplified through Chaps. 6 and 7 is continued. In particular these achievements are put together with perspectives, thus bridging the gap from: "What has been learned?" to "What can be learned?"

An appendix to the book summarizes numerical issues, which apply to the implementation of diffuse interface models. In particular, some of the latest high performance implementations are described. For the reader interested in his own application this provides information for the step from model to simulation. Moreover it gives him a feeling what simulations are possible from point of view of computational capacity.

2. What Is an Interface?
Interfaces in Materials Science and Beyond

If we think of an interface in materials science, then obviously this term applies to a boundary between two different phases of solid matter. In this sense we might distinguish:

- *liquid - liquid,*
- *liquid - solid,*
- *solid - solid,*
- *solid - vapor* and
- *liquid - vapor* interfaces.

It is relatively easy to agree on a common view for fluid interfaces, for which any physical quantity has a well defined smooth average profile. The situation is somewhat more complicated for solid interfaces, for which the bulk solid is modulated with the lattice periodicity. As a consequence one may view interfaces between two phases, one of them solid, in two ways:

1. First, viewed on an *atomic scale*, the position of every atom is specified. This leads to a description in terms of a terrace–step–kink model.
2. Second, viewed on a *coarse grained scale*, every property is averaged over a finite volume in such a way as to have constant averages in the bulk. These coarse grained quantities are then subject of a *continuum* description in much the same way as fluid interfaces.

Throughout this book I will choose the second approach. Then the notion of *surface excess* becomes essential to quantify extensive variables in the interface region. This term dates back to the work of Gibbs. For its explanation I start with a planar interface between two phases 1 and 2 located near the plane $x = 0$ (see Fig. 2.1). An arbitrary extensive quantity has a profile $\phi(x)$, where ϕ is the density per unit volume. In Gibbs' description, one deliberately ignores details of that profile. Rather one defines an integrated *surface excess* ϕ_s through the following construction: First choose an arbitrary dividing surface $x = \zeta$ somewhere in the interfacial region as indicated in Fig. 2.1a. Next extrapolate the bulk values ϕ_1 and ϕ_2 up to ζ. Then the surface excess ϕ_s is defined by

$$\phi_s = \int_{x_1}^{x_2} \phi(x)dx - \phi_2(x_2 - \zeta) - \phi_1(\zeta - x_1) , \qquad (2.1)$$

where x_1 and x_2 are well inside each phase (Fig. 2.1a).

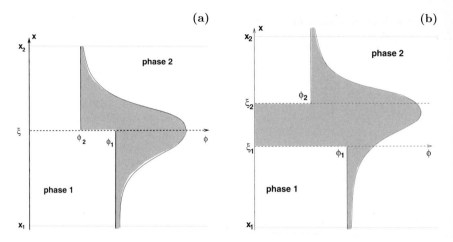

Fig. 2.1. Definition of surface excess ϕ_s (a) for a surface of *zero volume*, i.e. a *sharp interface*, as well as (b) for a finite surface, i.e. a *diffuse interface*.

It is obvious that the value of ϕ_s is dependent on the precise choice of ζ. If ζ changes by $\Delta\zeta$, then ϕ_s changes by

$$\Delta\phi_s = \Delta\zeta(\phi_2 - \phi_1) . \tag{2.2}$$

Only for quantities such that $\phi_1 = \phi_2$ the surface excess is defined unambiguously. In general any result referring to the surface must be invariant upon a change of ζ.

Equation (2.1) employs a single dividing surface, i.e. a surface of *zero volume*. It is often more convenient to use two distinct dividing surfaces ζ_1 and ζ_2. Then a definition of surface excess analogous to the one above results in the following formulation:

$$\phi_s = \int_{x_1}^{x_2} \phi(x)dx - \phi_2(x_2 - \zeta_2) - \phi_1(\zeta_1 - x_1) . \tag{2.3}$$

The underlying construction is depicted in Fig. 2.1b. For this case a change of ζ_1 and ζ_2 alters the surface excess ϕ_s in the following manner:

$$\Delta\phi_s = \phi_2\Delta\zeta_2 - \phi_1\Delta\zeta_1 . \tag{2.4}$$

In definition (2.4) the interface has a finite 'thickness' $l_s = \zeta_2 - \zeta_1$, i.e. it is *diffuse*. In this sense it is possible to quantify extensive variables in an interfacial region based on Gibbs' definition of surface excess for both, sharp as well as diffuse interfaces. Moreover, both interface types allow us to establish thermodynamic relations between those quantities dealing with mechanical or chemical equilibrium. The latter governs transfer of matter from one phase to the other and thus is essential for interfacial growth. Classically formulating chemical equilibrium results in the well known Gibbs–Duhem relation or the Gibbs–Thomson relation for liquid–liquid respective liquid–solid interfaces. For a minimal model describing the growth of a phase-separating interface only two additional physical requirements have to be met:

1. First, at the interface itself we would also require conservation of mass and energy to be satisfied.
2. Second, within the bulk at least the transport equation for the physical field governing the growth process would have to be taken into account.

Indeed for dendritic growth, which can be regarded as a paradigmatic problem of interfacial growth ever since Kepler [178], the Gibbs–Thomson relation together with an interfacial boundary condition for the conservation of mass and energy (Stefan condition) as well as an diffusion equation for heat transport in the bulk is sufficient to describe the growth process. Thus the minimal model for dendritic growth reads:

- **Bulk**

$$\frac{\partial T'}{\partial t} = D \, \nabla^2 T' \ . \tag{2.5}$$

- **Interface**
  Gibbs–Thomson relation

$$T' = T'_M \left(1 - \frac{\gamma}{L} \kappa \right) \ , \tag{2.6}$$

  Stefan condition

$$L\mathbf{v} \cdot \mathbf{n} = [(k\nabla T')_s - (k\nabla T')_l] \cdot \mathbf{n} \ . \tag{2.7}$$

Here equal diffusion constants D for the transport of heat in the liquid and in the solid phase have been assumed. Moreover, $k = Dc_p$ where c_p the specific heat per unit volume, which is assumed to be equal in both phases as well. T' is the temperature field, which governs the growth process. T'_M denotes the crystallization temperature of the planar interface, L the latent heat per unit volume of solid, γ the anisotropic liquid–solid surface tension, and κ the local curvature of the interface[1]. One assumes a fourfold symmetry[2] of the crystal, so that the surface tension can be written as

$$\gamma = \gamma_o a(\theta) = \gamma_o(1 - \beta \cos 4\theta) \ , \tag{2.8}$$

---

[1] The prime refers to the fact that at this point the temperature field is given in dimensional units.

[2] This multiple symmetry is a reminiscence of the underlying crystalline structure: Even though I'm only concerned with continuum descriptions of interfaces, the crystalline grid results in an orientation dependence of surface tension along the continuous interface also on the coarse grained scale. The precise dependence on orientation can be obtained via a Wulff construction [300]. According to such a construction an underlying fcc-grid gives rise to a fourfold symmetry.

where β is the anisotropy factor and θ the angle between the normal to the interface and the direction of propagation ($\cos\theta = \mathbf{n} \cdot \mathbf{e_x}$). The left-hand side of formula (2.7) constitutes the latent heat release per unit volume of solid. Its right-hand side accounts for the total energy flux away from the interface to both sides of the interface due to heat conduction.

Equations (2.5)–(2.7) constitute a sharp interface model for dendritic growth. Even though simple in nature it is in good agreement with experimental data for single dendrites grown into transparent materials[3] [128]. In this sense it can be understood as an accepted, a validated model.

Extending the model only slightly allows us to study directional solidification within a sharp interface formulation. This is particularly convenient for a validation through comparison to experiments, since directional solidification experiments can be carried out in a thin-film geometry for samples as thin as 12 μm. Using for example CBr$_4$ + 8 % C$_2$Cl$_6$ with diffusion length $\sim$ 50 μm [5] the relevant dynamics of the experiment takes place in the 2D plane. An experimental set-up underlying this type of directional solidification experiments is depicted in Fig. 2.2. The set-up specified here was used by Akamatsu and Faivre to study the doublon–dendrite transition in directional solidification [5]. Pictures of the doublon morphology as obtained in their experiments are displayed in Fig. 2.3. Simulations by Ihle revealed their relation to a minimal sharp interface model of directional solidification, which in terms of a dimensionless diffusion field u explained further in (2.15)–(2.17) reads:

- **Bulk**

$$\frac{\partial}{\partial t}u = D\nabla^2 u, \tag{2.9}$$

- **Interface**
  Gibbs–Thomson relation

$$u|_{\text{Int}} = 1 - d(\theta)\kappa - \frac{\xi - Vt}{l_T} - \beta_{\text{kin}}(\theta)v_n , \tag{2.10}$$

Stefan condition

$$((1 - K)u_{\text{Int}} + K)\mathbf{v} \cdot \mathbf{n} = -D\nabla u_{\text{Int}} \cdot \mathbf{n} . \tag{2.11}$$

Here $\xi = \xi(y) := 0$ denotes the x-position of the interface and $d(\theta) = d_0(1 - \beta\cos 4\theta)$, where d_0 is the capillary length which is related to the surface tension of (2.8) via

$$d_0 \equiv \frac{\gamma c_p T'_M}{L^2} . \tag{2.12}$$

Thus it inherits the fourfold symmetry of γ:

---

[3] Systematic deviations from the theory at small undercoolings will be discussed in detail in Chap. 6.

$$d(\theta) \equiv \left(\gamma(\theta) + \frac{\partial^2 \gamma(\theta)}{\partial \theta^2} \right) \frac{c_p T_M'}{L^2} . \qquad (2.13)$$

The fourth term on the right hand side of the Gibbs–Thomson relation, i.e. $\beta_{\text{kin}}(\theta)v_n$, denotes deviations from local thermodynamic equilibrium. It becomes relevant, if the interface is growing rapidly and the idea of an instantaneous relaxation of the physical fields to the position of the interface does not hold any longer. Just as capillary effects kinetic effects can be anisotropic: $\beta_{\text{kin}}(\theta) = \beta_{\text{kin}_0}(1 - \beta_{\text{kin}_4}\cos 4\theta)$, where the kinetic anisotropy factor β_{kin_4} does not necessarily have to be equal to the crystalline anisotropy factor β. Moreover,

$$l_T := \frac{m_l \Delta C}{G} , \qquad (2.14)$$

where m_l is the so-called liquidus slope, ΔC the miscibility gap and G the temperature gradient. K denotes a so-called *distribution coefficient*.

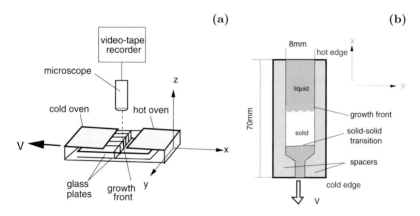

Fig. 2.2. This experimental set-up was used by Akamatsu and Faivre for their experiments on the directional solidification of CBr$_4$ + 8 % C$_2$Cl$_6$ thin-films [5]. (a) Principle experimental set-up. (b) More specific sketch of the sample itself. x is the direction, in which the sample is pulled with constant velocity V.

The picture underlying that model is that of a growth process driven by a solute field rather than a temperature field. This implies that the experiments are no longer done with pure materials but rather with alloys. Moreover, within a directional solidification experiment the sample containing the alloy is moved with constant velocity V within an externally opposed temperature gradient such that the position of the interface is kept fixed. This geometry is depicted in Fig. 2.2 on the right. If the heat conductivity of liquid and solid phase are approximately equal, then the temperature profile between the hot edge and the cold edge (see Fig. 2.2) can well be assumed as:

$$T = T_0 + Gx \ . \tag{2.15}$$

By an appropriate choice of the temperature T_0 it is possible to keep the position of the interface at $x = 0$. The diffusive transport equation, which remains to be solved, is that of the solute field C. Formulating solute diffusion one has to take into account, that the value of the solute field depends on temperature. Figure 2.4 depicts the dependence of C on T. Here liquidus line, the slope of which appears as m_l within (2.14), solidus line with slope m_s and the miscibility gap ΔC are displayed. The distribution coefficient K appearing within (2.11) can now be understood as the ratio

$$K = \frac{C_\infty}{C_\infty + \Delta C} \ , \tag{2.16}$$

where C_∞ denotes the concentration of the solute field at $T = T_0$ (see Fig. 2.4). Moreover, the governing field u, in terms of which model equations (2.9)–(2.11) are written, reads

$$u(x, y, t) = \frac{C(x, y, t) - C_\infty}{\Delta C} \ . \tag{2.17}$$

Apart from the experiments by Akamatsu and Faivre quantitative validation of sharp interface modeling for directional solidification was obtained e.g. through experiments by Bechhoefer *et al.* [25, 258] as well as Trivedi *et al.* [275]. These experiments were concerned with the so called *limit of absolute stability* [223]. The limit of absolute stability is a maximal velocity of a directional solidifying front in a parameter regime, where it is stable whatever the value of the thermal gradient is. Theoretically it was predicted to be $\frac{D}{d_0 K}$ [223]. Indeed this was observed in the experiments by Bechhoefer *et al.* on the liquid-to-nematic transition of 8CB [25, 258] as well as by Trivedi *et al.* on distillated CBr$_4$ [275]. Other evidence for a validation of sharp interface models from directional solidification experiments centered around the issues of shape and velocity selection, e.g. [27, 90, 112, 249, 260, 274].

To summarize the experimental evidence at this point, one may state that there is a long tradition in sharp interface modeling of solidifying phase–boundaries and relating these models to experiments. The same is true for solid–vapor or solid–liquid interfaces subject of epitaxial growth (see, e.g., [288] and references within). Thus it seems justified to construct such kind of models simply from formulating the governing equations for transport in the bulk phases coupled to mass and energy conservation (Stefan condition) as well as the requirement of local thermodynamic equilibrium (Gibbs–Thomson condition) at the interface.

From a mathematical point of view these problems constitute a class of so-called *moving boundary problems*. Usually analytical treatment of these type of problems is very restricted. Thus it seems natural to search for adequate numerical treatment. However, such numerical treatment proofs difficult, as well.

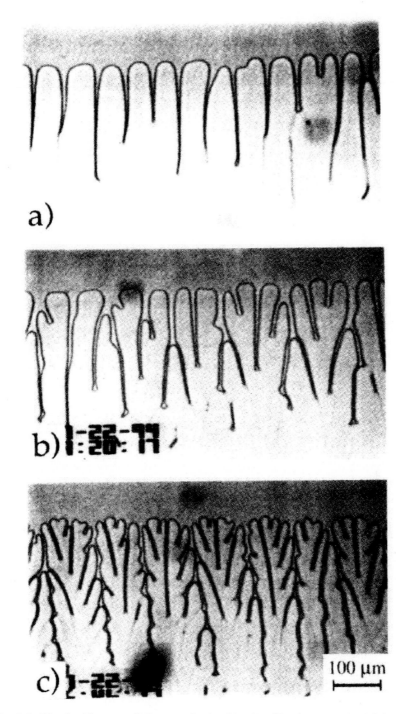

Fig. 2.3. The doublon morphology as obtained in thin-film directional solidification with $CBr_4 + 8\ \%\ C_2Cl_6$ by Akamatsu and Faivre. Picture taken from [5].

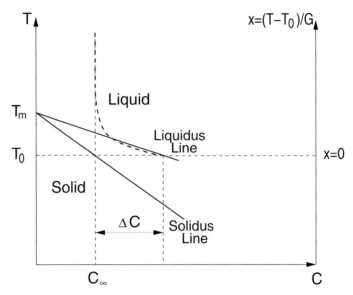

Fig. 2.4. Part of the phase diagram of a binary alloy. Liquidus and solidus curve are depicted just as well as the miscibility gap ΔC and the alloy melting temperature T_m. Via an appropriate choice of the temperature T_0 it is possible to keep the position of the interface at $x = 0$ (see (2.15)). $C_\infty = C^S(T_0)$ refers to the concentration of solute in the solid phase at $T = T_0$. Due to (2.15) the temperature scale given on the left vertical axis of the diagram corresponds directly to the in growth direction (i.e. the x-coordinate) of the sample as indicated by the second vertical axis of the diagram to the right.

Ideally one would like to simulate the effect of the boundary by treating it explicitly, i.e. with no smearing of information at the interface resulting in numerical diffusion. For fixed boundaries, even if very complex in shape, techniques for such an explicit treatment include block–structured domain decomposition [256], overset meshes [156, 264] or unstructured boundary–conforming curvilinear grids [286] to discretize the domain. For moving boundaries, on the other hand, fixed grid techniques are employed, in which the computations are performed on the fixed grid while at the same time the interface is tracked explicitly as independent curve. These approaches constitute mixed Eulerian–Lagrangian methods. Examples of such approaches are the immersed boundary technique [157, 233, 280, 285], cut-cell type approaches [230, 241, 257, 278, 279], the immersed interface method [206] and the fictitious domain methods [132].

The complications arising from such a mixed Eulerian–Lagrangian method are twofold:

1. First, it implies two distinct schemes of discretization, namely that for the moving interface opposed to the fixed grid employed for the transport equations. However, using the Gibbs–Thomson relation properly as

boundary condition for the heat transport field requires to interpolate between these two discretization schemes and thus brings about unavoidable interpolation errors. Moreover, the Stefan condition requires to determine normal gradients away from the interface onto the fixed grid, which again involves interpolations. Thus ensuring numerical stability and adequate accuracy gets expensive in terms of CPU time.

2. Second, for Eulerian–Lagrangian schemes the number of discretization points of the scheme varies from time step to time step. As a result these approaches are difficult to parallelize efficiently. However, having in mind applications to real phenomena posing open questions such as 3D systems including more than just diffusive transport dynamics, efficient implementation becomes essential to carry out the parameter studies necessary to gain new physical insight.

Thus if it were only for numerics, a *diffuse interface approach* in materials science would seem to be a natural advance to treat interfacial dynamics numerically, since it overcame item 1 as well as item 2 of the enumeration above. In this sense one would employ diffuse interface models along the lines of the first philosophy I pointed out in the introductory chapter, i.e. merely as a computational trick. Other philosophies motivating diffuse interface theory also from the point of view of new perspectives with respect to the modeling itself have been discussed within the introductory chapter as well. Their impact will become clearer in Chaps. 6 and 7 of this book, being concerned with phenomena, for which particularly the diffuse interface approach provides a way to gain new insight. Moreover I will summarize within the final chapter of the book in which way new perspectives for modeling as well as numerics open up new chances for gaining insight in evolving phase boundaries.

In this context phase boundaries are more than the interfaces known from materials science. Talking of *interfaces in a broader sense* one could include evolving boundaries from e.g. combustion, image processing, computer vision, control theory, seismology and computer aided design, as well. For some of these problems the moving interface seems virtual and emerges only after some sort of transformation. For example for the problem of shape-from-shading in computer vision, one has to map the image segmentation problem onto a moving interface version of active contours. However, for all of these examples the key ingredient to construct a diffuse interface model is to rethink the Lagrangian geometric perspective and replace it with an Eulerian, partial differential equation. Having accomplished that task, the gain is advanced numerical treatment as well as new perspectives for modeling just as in the case of materials science.

In many aspects the diffuse interface approach resembles features of the level set method. Nevertheless there are clear differences, as well. To conclude this chapter I will therefore contrast the two methods for applications in materials science:

The level set method was first introduced by Osher and Sethian [228]. Since then level set algorithms have successfully been applied to a wide variety of problems [66, 70, 114, 148, 215, 266, 302]. The fundamental idea behind the level set method is the representation of the interface Γ_{ij} by the help of a zero level set function $\phi(\mathbf{x}, t)$, i.e.:

$$\Gamma_{ij} = \{\mathbf{x} : \phi(\mathbf{x}, t) = 0\} . \tag{2.18}$$

Given a velocity field $\mathbf{v}$, one can analyze the dynamics of Γ_{ij} by relating it to the motion of the zero level set of ϕ. The partial differential equation describing the temporal evolution of ϕ such that the level sets move with $\mathbf{v}$ is

$$\frac{\partial \phi}{\partial t} + \mathbf{v} \cdot \nabla \phi = 0 . \tag{2.19}$$

Since the normal vector $\mathbf{n}$ can be written in terms of ϕ as $\mathbf{n} = \frac{\nabla \phi}{|\nabla \phi|}$ and since further $\mathbf{v} = v_n \mathbf{n}$, (2.19) is equivalent to the level set equation:

$$\frac{\partial \phi}{\partial t} + v_n \cdot |\nabla \phi| = 0 . \tag{2.20}$$

The difficult issue when employing the level set approach is the derivation of the velocity function $\mathbf{v}$ and its extension to the complete domain of the level set function ϕ as required to solve (2.20). The extension of $\mathbf{v}$ to the complete domain can be achieved via the so-called *Fast Marching Method* [253], which results in a first-order accurate solution only, if linear triangular elements are used. Higher order elements enable higher order solutions, however at a much higher computational price.

Moreover the boundary conditions have to be evaluated at the boundary Γ_{ij}, now given by $\Gamma_{ij} = \{\mathbf{x} : \phi(\mathbf{x}, t) = 0\}$. Γ_{ij} does not necessarily coincide with the grid points of the underlying Eulerian discretization scheme. As a consequence the boundary conditions cannot be resolved on exactly this grid. In [71] this problem is treated via interpolations, which lead to a smearing of information over the interface and limit the capability of this method to take into account precise kinetics of an interface. In [72] an alternative way is chosen, i.e. the problem is met by employing a finite element mesh with linear triangular elements, whose nodes are placed exactly on the boundary. Obviously there is a price to pay for this as well, namely tedious mesh adaption.

The basic difference between the level set method and the phase-field method is that the latter method depends on a small parameter for the interface thickness. Thus the interface is not a sharp interface as in the level set case, and boundary conditions have to be fulfilled only asymptotically for vanishing interface thickness. Obviously this restricts the operation of this method to interface thicknesses proportional to the grid size. As a result it used to be hard to employ the phase-field method in a quantitative manner, since it proved to be computationally demanding to choose the interface thickness small enough to resolve the desired sharp interface limit. In the context of dendritic growth efforts to overcome this difficulty resulted in the so-called *thin interface* analysis [168]. This *thin interface* analysis also improved the capability of the phase-field method to model kinetic effects in detail considerably. I will discuss it in more detail in Chap. 5.

For further information on the level set method the interested reader is referred to [254], which is a comprehensive and self-contained introduction to that field.

3. Equilibrium Thermodynamics of Multiphase Systems: Thermodynamic Potentials and Phase Diagrams

A first step when entering the field of thermodynamics certainly is concerned with equilibrium thermodynamics, i.e. with gaining an understanding about how a system can be characterized by macroscopic *state variables* and *potentials* and how those can be organized into *phase diagrams*[1] for systems at equilibrium. *Thermodynamic equilibrium* is reached, if the state of a system remains constant, unless variation of subsidiary conditions triggers a change of governing variables. Thus a definition of thermodynamic equilibrium always comes along with a set of well defined rules (the subsidiary conditions), which tell us, which (depending) variables can change freely[2] and which variables are the state governing ones, which will have to be kept constant. A thermodynamic state is described fully by a minimal set of governing variables, which are independent of each other. The number of those independent governing variables is system dependent.

To describe a thermodynamic equilibrium situation mathematically, the notion of a *thermodynamic potential* is essential, since it allows us to define the equilibrium state as the one, for which the governing thermodynamic potential becomes extremal. The basic thermodynamic functions to consider within this context are the Helmholtz free energy A and the Gibbs free energy G. One considers them as "thermodynamic potentials", if they appear as functions of their natural variables. For the Gibbs free energy G the natural variables are the pressure P, the temperature T, and the molar number n of a system. For the Helmholtz free energy A, on the other hand, the system's volume V replaces the pressure P in the set of natural variables. With these respective dependencies G and A allow for a complete description of a ther-

---

[1] A phase diagram depicts the two phases solid and liquid in dependence on the three variables temperature, concentration and pressure - thus a full *phase diagram* should be three dimensional. However, since approximately constant pressure is assumed, here the diagram turns out to be a 2D plot just as in Fig. 3.1. Within the diagram equilibrium phases are denoted by "S" (solid) and "L" (liquid). Moreover, in the particular diagram given by Fig. 3.1, there is exactly one third domain in the diagram referring to coexistence of liquid and solid phase ("S+L"). The curves separating the coexistence domain from the solid and liquid domain, respectively, are the solidus and liquidus curve introduced in Fig. 2.4.

[2] and in fact will change during the evolution of the system towards thermodynamic equilibrium

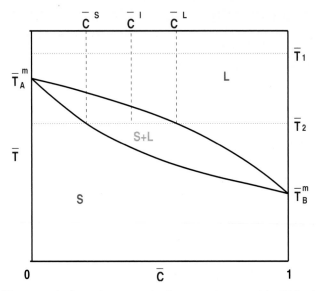

Fig. 3.1. The typical phase diagram of a binary mixture with Gibbs free energies of both phases as given in Fig. 3.2 for $\overline{T}_1$ and Fig. 3.3 for $\overline{T}_2$.

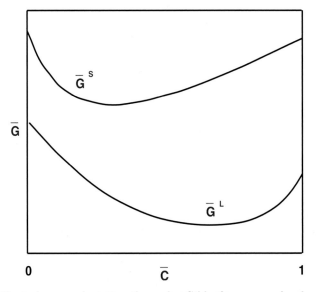

Fig. 3.2. Typical curves depicting the molar Gibbs free energy for the solid phase $\overline{G}^S$ and the liquid phase $\overline{G}^L$ of a binary mixture at a specific temperature $\overline{T}_1$. Logarithmic contributions due to the mixture entropy are neglected. On the $\overline{C}$-axis the relative portion of species B is plotted. Here the liquid phase is the stable one for all mixture ratios (compare to Fig. 3.3).

modynamic system in the sense that the physical variables for all well defined thermodynamic states can be determined from them via extremal principles.

In the following I will first concentrate on the Gibbs free energy G and its meaning for simple "two-phase" systems. This concept of G is an essential background to describe the dynamics of phase separating interfaces later. I will then specify some relevant Gibbs energies with respect to special classes of model systems and also explain special phase diagram approaches to those systems. This will allow me to elucidate the discrepancy of a "locally non-conserving-variable-approach" and a "locally conserving-variable-approach" with respect to the concept of *thermodynamic consistency* as one of the key ideas for the remainder of this work.

The following concepts can be found in more detail within elementary thermodynamic textbooks as e.g. [4, 184]. The 'short-cut' introduction presented here follows mainly [2].

3.1 Calculating Phase Diagrams from Energy Functionals

As mentioned in the previous section, the equilibrium configuration of a thermodynamic system is characterized by an extremum of the relevant thermodynamic potential. For binary mixtures, for which two phases "S" (solid) and "L" (liquid) can exist, the differential of the Gibbs free energy for each of the two phases is given by:

$$d\overline{\mathcal{G}} = d\overline{\mathcal{G}}^S + d\overline{\mathcal{G}}^L , \qquad (3.1)$$

where

$$d\overline{\mathcal{G}}^\pi = \overline{V}_\pi dP - \overline{S}_\pi d\overline{T} + \sum_{k \in \{A,B\}} \overline{\mu}_{k,\pi} d\overline{n}_{k,\pi} \qquad ; \qquad \pi \in \{L, S\} . \quad (3.2)$$

In the above equations $\overline{P}$ denotes the pressure of the system and $\overline{T}$ its temperature. These are the state governing variables. Thus they are not allowed to depend on the phase π. $\overline{V}_\pi$ and $\overline{S}_\pi$ are the volume and the entropy of phase π, respectively. $\overline{n}_{k,\pi}$ is the molar number of component $k \in \{A, B\}$ in phase π. The chemical potentials are defined by

$$\overline{\mu}_{k,\pi} = \frac{\partial \overline{\mathcal{G}}^\pi}{\partial \overline{n}_{k,\pi}} = \frac{\partial \overline{G}^\pi}{\partial \overline{C}_{k,\pi}} , \qquad (3.3)$$

with the molar Gibbs free energy $\overline{\mathcal{G}}^\pi$ and a relative portion of $\overline{C}_{k,\pi}$ of component k in phase π. At this point surface effects at the phase-separating interface, which will be an important ingredient of succeeding chapters dealing with the dynamics of the systems, are neglected. Assuming approximately constant pressure, the $\overline{P}$-dependence in (3.2) can be omitted, as well.

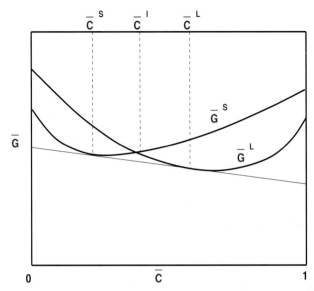

Fig. 3.3. Same as in Fig. 3.2, however at a different temperature $\overline{T}_2$, which is smaller than the temperature $\overline{T}_1$ chosen for Fig. 3.2. As a result the energy curve displays an equilibrium configuration at which both phases coexist.

In Fig. 3.2 typical forms of $\overline{G}^{L/S}$ are displayed as a function of the relative portion of species B in the respective phases for a fixed temperature $\overline{T}_1$. As proven in [142] the molar Gibbs free energy always takes a local maximum for mixings 0 and 1. Between these two points the graph is a convex curve. In Fig. 3.2 $\overline{G}^S$ is larger than $\overline{G}^L$ for all mixture ratios. Thus the liquid phase is the stable phase for arbitrary concentrations. Reducing the temperature, energetic configurations will be changed in favor of the solid phase. At a certain temperature $\overline{T}_A^m < \overline{T}_1$, the melting temperature of A, both curves will coincide at a concentration $\overline{C} = 0$. Reducing the temperature further to $\overline{T}_2$ results in an intersection of curves and thus a coexistence of phases (Fig. 3.3). The equilibrium configuration can be obtained from the following consideration [294]: The identities

$$\overline{\mu}_{A,\pi} = \overline{G}^{\pi}(\overline{C}_{B,\pi}) - \overline{C}_{B,\pi}\frac{\partial \overline{G}^{\pi}(\overline{C}_{B,\pi})}{\partial \overline{C}_{B,\pi}} \tag{3.4}$$

$$\overline{\mu}_{B,\pi} = \overline{G}^{\pi}(\overline{C}_{B,\pi}) + (1 - \overline{C}_{B,\pi})\frac{\partial \overline{G}^{\pi}(\overline{C}_{B,\pi})}{\partial \overline{C}_{B,\pi}} \tag{3.5}$$

reveal that the chemical potential for a concentration $\overline{C}_{B,\pi}$ is determined by the value of the curve tangential to $\overline{G}_{\pi}$ in $\overline{C}_{B,\pi}$. On the other hand, because of conservation of species as implied by (3.1) and (3.2), one obtains the equilibrium condition

$$\overline{\mu}_{k,L} = \overline{\mu}_{k,S} \,, \tag{3.6}$$

which implies that at equilibrium the two curves tangential to $\overline{G}^L$ and $\overline{G}^S$, respectively, result in identical values for $\overline{C}_B = 0$ and $\overline{C}_B = 1$. This condition is fulfilled by a double–tangent curve, which joint points with $\overline{G}^L$ and $\overline{G}^S$ yield the equilibrium concentrations in liquid ($\overline{C}^L$) as well as solid phase ($\overline{C}^S$) (see Fig. 3.3).

Further reduction of the temperature results in changes of this value towards higher concentrations. Finally, if the melting temperature of component B is reached, the solid phase is stable for all mixture ratios. If one plots the equilibrium concentration values versus temperature, one obtains the phase diagram of two completely miscible species (Fig. 3.1). There is a two-phase state "S+L", for which the system minimizes the value of $\overline{\mathcal{G}}$ through the coexistence of solid and liquid phase. The relative amount of the two phases at equilibrium is given by the ratio $(\overline{C}^L - \overline{C}^I)/(\overline{C}^I - \overline{C}^S)$, where $\overline{C}^I$ is the overall fraction of species B in the mixture. A phase diagram of this kind is for example found for CuNi-mixtures [198].

In general the exact form of a phase diagram depends on the two functions $\overline{G}^S$ and $\overline{G}^L$. If, for example, both of them take a minimum at nearly identical mixture ratios, then the curve for $\overline{G}^L$ displays a stronger curvature than the one for $\overline{G}^S$. As a result, a further increase in temperature[3] results in an intersection of the two curves at a concentration close to the two minima. A phase diagram different from the one depicted in Fig. 3.1 in the sense that two distinct (but not disjunct) coexistence regimes arise, is the result. Such a kind of phase diagram can for example be found for CsK-mixtures [198]. For a K concentration of 50.5 % solidus and liquidus slopes coincide. At this point both curves are at a minimum (Fig. 3.4). The "real" phase diagram displayed in Fig. 3.4 depicts the phase boundary of a CsK_2-phase (lower right corner of the figure), as well.

This method to calculate the phase diagram from energy functionals is not restricted to a system of two components, i.e. a binary mixture. By introduction of suitable potentials even the complex phase diagrams of multicomponent–multiphase systems can be calculated as well as validated from experimental findings [62].

3.2 Abstracted Phase Diagrams

If one focuses ones main interest on the physical processes underlying the dynamics of two-phase systems and limits ones attention to certain mixture ratios, then simplified phase diagrams can be developed. They can be understood as abstractions of basic types of phase diagrams as, for example, the one depicted previously in Fig. 3.1.

---

[3] The starting point for this consideration is a temperature, at which the system is solid for all concentrations.

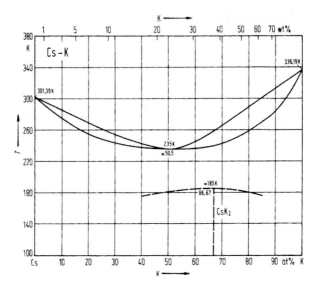

Fig. 3.4. "Real" phase diagram of a CsK-mixture [198]. Compared to the phase diagram depicted in Fig. 3.1 this displays a second, qualitatively different type of diagram found for two-phase systems containing two distinct but not disjunct coexistence regimes.

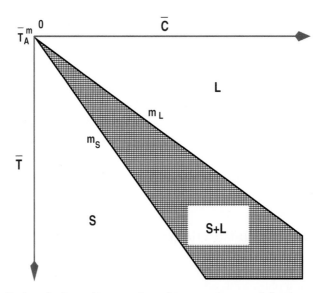

Fig. 3.5. Reduced phase diagram for a binary mixture of dominant component A and only small portions of B (i.e. weakly contaminated melts). The coexistence domain is hatched. It is bounded by the solidus curve with slope $-m_S$ and the liquidus curve with slope $-m_L$. These two curves intersect at the melting temperature $\overline{T}_A^m$.

There are many questions concerned with systems, which involve a dominant phase A mixed with only a small amount of a second species B. The prototype is a weakly contaminated melt. In this case it is important to know the precise configuration of the phase diagram only for a small interval of concentration values. Then it is justified to expand solidus and liquidus lines around the melting point $\overline{T}_A^m$, so that the two-phase domain is bounded by solidus and liquidus curve with constant slopes $-m_L$ and $-m_S$, respectively. Here $-m_L$ and $-m_S$ are both larger than 0 (Fig. 3.5). Reduced phase diagrams of this kind have extensively been used for theoretical and analytical investigations of solidification phenomena in binary mixtures in the past [28, 53, 65, 80, 81, 82, 125, 162, 163, 164, 199, 202].

If for a phase diagram as depicted by Fig. 3.1 the interesting concentration of species B turns out to be in the range of $\overline{C} = 0.25$ to $\overline{C} = 0.75$, Fig. 3.5 is no longer a suitable abstraction. Rather liquidus and solidus curve can be assumed to be approximately parallel, i.e. $m = m_L = m_S$. This type of reduced phase diagram is depicted in Fig. 3.6. It is characterized by a miscibility gap, which is independent of the concentrations and implies invariance of translation along the solidus and the liquidus curve. As a result, deriving the energetic potentials to obtain dynamic equations subsequently is particularly simple. At first sight, a severe short-coming of this type of diagram seems to be the absence of a well defined melting temperature[4]. However, this does not have any consequences for the derivation of the evolution equations of the system, since the melting temperature[5] is only relevant to determine the capillary length. In Sect. A.1 I will show that the capillary length is still well defined for systems represented by Fig. 3.6 and that it can be obtained in an alternative manner. Thus the absence of a precise melting point does not interfere with the derivation of model equations, which are meant to display the relevant physical effects. In this sense the reduced phase diagram Fig. 3.6 itself can be understood as comprising all information necessary to derive the dynamic equations of the system. Because of its convenient behavior with respect to this point, I will present the mathematical formalism of the succeeding chapters based on it as representative for the underlying thermodynamics. Usually the formalism can be extended to cover systems described by a reduced phase diagram as in Fig. 3.5, as well.

The phase diagram of Fig. 3.6 is characterized by four parameters: M_C, M_T, C_0 and T_0. They can be transformed into the two dimensionless variables

$$C \equiv \frac{\overline{C} - \overline{C}_0}{M_C} \qquad ; \qquad T \equiv \frac{\overline{T} - \overline{T}_0}{M_T} . \tag{3.7}$$

T and C allow us to construct a parameter-free model phase diagram. Now the miscibility gap as well as the temperature difference between liquidus and solidus line is 1, moreover $m_S = m_L = 1$. In the following section I will

---

[4] Since there is no intersection of liquidus and solidus curve.
[5] if it is available as in Fig. 3.5

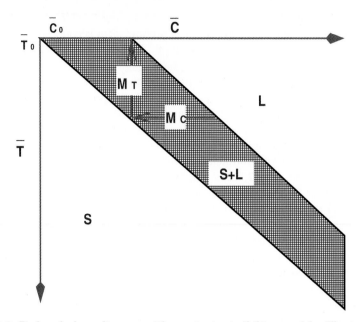

Fig. 3.6. Reduced phase diagram with constant miscibility gap M_C. The temperature difference M_T of solidus and liquidus curve is constant. Therefore both curves display the same slope $-m$. The reference point $(\overline{C}_0, \overline{T}_0)$ can be chosen arbitrarily.

discuss the derivation of the Gibbs free energy in the liquid as well as the solid phase for such a parameter-free diagram.

3.2.1 Constructing the Gibbs Free Energies

Close to thermodynamic equilibrium G^L and G^S can be expanded as:

$$G^L(C,T) = A_1(C - A_2)^2 + A_3(C - A_4)^2 + A_5CT \qquad (3.8)$$
$$G^S(C,T) = B_1(C - B_2)^2 + B_3(T - B_4)^2 + B_5CT + B_6 , \qquad (3.9)$$

with expansion parameters A_i, B_i. For an arbitrary temperature T the equilibrium concentration in liquid and solid is given by

$$C^L(T) = 1 - T \qquad ; \qquad C^S(T) = -T . \qquad (3.10)$$

The equilibrium conditions following from the double–tangent construction at G^L and G^S (see (3.4)–(3.6)) read

$$\left.\frac{\partial G^L(C,T)}{\partial C}\right|_{C=C^L} = \left.\frac{\partial G^S(C,T)}{\partial C}\right|_{C=C^S} \qquad (3.11)$$

$$G^L(C^L,T) = G^S(C^S,T) + \left.\frac{\partial G^S(C,T)}{\partial C}\right|_{C=C^S} (C^L - C^S) . \quad (3.12)$$

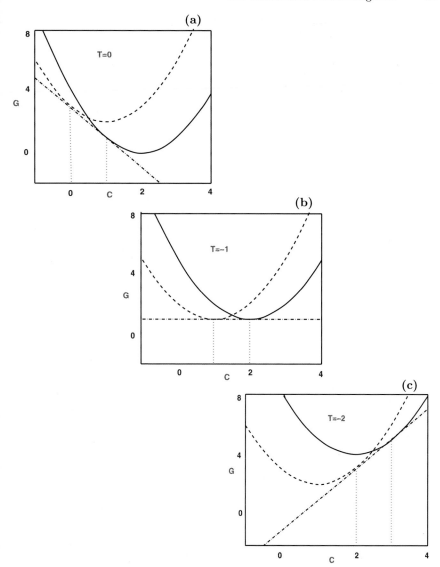

Fig. 3.7. For three different temperatures the Gibbs free energy of liquid phase (solid line) and solid phase (dashed line) are plotted: (a) $T = 0$, (b) $T = -1$ and (c) $T = -2$. The energies follow from (3.21). The respective equilibrium concentrations following from the double–tangent construction are indicated.

They have to be fulfilled for arbitrary T and thus at any order of T, so that the A_i and B_i obey the following relations:

$$A_1(1 - A_2) = -B_1 B_2 , \tag{3.13}$$
$$A_1(1 - A_2)^2 + A_3 A_4^2 = B_1 B_2(B_2 - 2) + B_3 B_4^2 + B_6 , \tag{3.14}$$
$$A_5 - 2A_1 = B_5 - 2B_1 , \tag{3.15}$$
$$2A_1(A_2 - 1) - 2A_3 A_4 + A_5 = 2B_1(B_2 - 1) - 2B_3 B_4^2 + B_5 , \tag{3.16}$$
$$A_1 + A_3 - A_5 = B_1 + B_3 - B_5 . \tag{3.17}$$

These can be transformed to

$$A_1 = \frac{A_5}{2} + B_1 - \frac{B_5}{2} \qquad A_2 = 1 + \frac{2B_1 B_2}{A_5 + 2B_1 - B_5} , \tag{3.18}$$

$$A_3 = \frac{A_5}{2} + B_3 - \frac{B_5}{2} \qquad A_4 = 1 + \frac{2B_1 + 2B_3 B_4 + A_5 - B_5}{A_5 + 2B_3 - B_5} , \tag{3.19}$$

$$B_6 = \frac{2B_1^2 B_2^2}{A_5 + 2B_1 - B_5} + \frac{(2B_1 + 2B_3 B_4 + A_5 - B_5)^2}{2(A_5 + 2B_3 - B_5)}$$
$$+ 2B_1 B_2 (1 - \frac{B_2}{2}) - B_3 B_4^2 . \tag{3.20}$$

A_5, B_1, B_2, B_3, B_4 and B_5 remain parameters one can choose freely. If the energy term does not contain any C–T coupling (i.e. $A_5 = B_5 = 0$), one obtains $B_1 = B_2 = B_3 = -B_4 = 1$ and thus particularly simple energy functionals in solid and liquid. They are visualized for three different values of $\overline{T}$ in Fig. 3.7. Formally they are given by:

$$G^L(C, T) = (C - 2)^2 + T^2 \tag{3.21}$$
$$G^S(C, T) = (C - 1)^2 + (T + 1)^2 + 1 . \tag{3.22}$$

3.2.2 Non-conserved Versus Conserved Variable Approach

In succeeding chapters of this text one of the key issues is how to derive a thermodynamically consistent phase-field model for the dynamics of the phase-separating interface based on a governing thermodynamic potential. A key element to accomplish this task is to substitute the dimensionless temperature T in (3.21) and (3.22) through the dimensionless energy density Q. The necessity to do so is twofold:

1. First, the formalism to derive an evolution equation from the governing potential is different depending on whether a variable is a conserved one or a non-conserved one. This will become clearer in Chap. 4, where this formalism is explained. As a consequence it is quite essential to distinguish strictly between locally conserving and locally non-conserving variables of the system (see (4.42) and its explanations).

2. Second, if evolution equations are derived from functionals, which contain the temperature as variable rather than the energy density, then the dynamic behavior is characterized by the feature, that changes to temperature cause changes in energy, as well. The thermodynamic potentials are no longer monotonically decreasing in time as could be proven by Penrose and Fife [231, 232]. A transition from temperature to inner energy as governing variable overcomes this problem. Thus it opens a way to derive thermodynamically consistent phase–field models.

If one replaces T by Q, one has to pay attention to the fact, that the double–tangent construction given by (3.11) and (3.12) is no longer suitable. A more general formula has to be obtained. To do so, one minimizes the energy functional with respect to the remaining independent system variables by at the same time ensuring the necessary conservation conditions for the functional. To obtain an adapted formula for the double–tangent construction one considers a closed binary system with species A and B and with fixed number of particles N_A and N_B ($N \equiv N_A + N_B$). Moreover, the total energy $E = E^S + E^L$ is fixed. E^S and E^L are contributions to the energy E from liquid and solid phase, respectively. In addition,

$$N_S \quad : \qquad \text{total number of particles in solid phase}$$
$$n \quad : \qquad \text{number of particles of species B in solid phase}$$
$$\hat{C}^S \quad : \qquad \text{concentration of species B in solid phase: } \hat{C}^S \equiv n/N$$
$$\hat{C}^L \quad : \qquad \text{concentration of species B in liquid phase: } \hat{C}^L \equiv \frac{N_B - n}{N - N_S}$$
$$\hat{Q}^S \quad : \qquad \text{energy per particle in solid phase: } \hat{Q}^S \equiv E^S/N_S$$
$$\hat{Q}^L \quad : \qquad \text{energy per particle in liquid phase: } \hat{Q}^S \equiv \frac{E - E^S}{N - N_S}.$$

Here variables $\hat{C}^S$, $\hat{C}^L$, ... have been introduced, which are identical with the variables C^S, C^L, ... except for a normalization factor.

At this point the configuration of the system is determined by the values of n, N^S and $\hat{Q}^S$, as well as N, N_B and E, which are external parameters. Minimizing

$$\hat{\mathcal{G}}(N, N_B, E, n, N^S, \hat{Q}^S) \equiv N_S \hat{G}^S(\hat{C}^S, \hat{Q}^S) + (N - N^S)\hat{G}^L(\hat{C}^L, \hat{Q}^L) \quad (3.23)$$

with respect to n, N^S and $\hat{Q}^S$ results in

$$\frac{\partial G^S(C^S, Q^S)}{\partial C^S} = \frac{\partial G^L(C^L, Q^L)}{\partial C^L} \tag{3.24}$$

$$\frac{\partial G^S(C^S, Q^S)}{\partial Q^S} = \frac{\partial G^L(C^L, Q^L)}{\partial Q^L} \tag{3.25}$$

$$G^S(C^S, Q^S) - G^L(C^L, Q^L) = C^S \frac{\partial G^S(C^S, Q^S)}{\partial C^S} - C^L \frac{\partial G^L(C^L, Q^L)}{\partial C^L}$$
$$+ Q^S \frac{\partial G^S(C^S, Q^S)}{\partial Q^S} - Q^L \frac{\partial G^L(C^L, Q^L)}{\partial Q^L}. \tag{3.26}$$

These conditions represent a set of possible surfaces tangential to G^L and G^S, which allow us – after fixing one variable, e.g. Q^S – to determine the equilibrium conditions of the remaining variables Q^L, C^L and C^S.

Once the relation between T and Q in both phases is fixed, the three equations (3.24)–(3.26) can be used to calculate the functions $G^L(C^L, Q^L)$ and $G^S(C^S, Q^S)$. For this it is convenient to transform the latent heat[6] $\overline{L}$ and the energy density $\overline{Q}$ to dimensionless variables as:

$$Q \equiv \frac{\overline{Q} - \overline{c}_p \overline{T}_0}{\overline{c}_p M_T} \qquad ; \qquad L \equiv \frac{\overline{L}}{\overline{c}_p M_T} \qquad (3.27)$$

(here $\overline{c}_p$ denotes the specific heat of the material). Afterwards the relation between dimensionless energy density and dimensionless temperature reads[7]:

$$T = \begin{cases} Q^S & : \text{ in the solid phase} \\ Q^L - 1 & : \text{ in the liquid phase} \end{cases} \qquad (3.28)$$

If the system is at thermodynamic equilibrium the relevant phase diagram has to fulfill

$$C^S = C^L - 1 \qquad ; \qquad Q^S = Q^L - 1 \qquad ; \qquad C^S = -Q^S. \quad (3.29)$$

If one determines[8] the parameters A_i and B_i such that (3.24)–(3.26) and (3.29) are satisfied, the thermodynamic functional reads

$$\begin{aligned} G^L(C, Q) &= (C-2)^2 + Q^2 \\ G^S(C, Q) &= (C-1)^2 + (Q+1)^2 . \end{aligned} \qquad (3.30)$$

These forms of the functional will be the basis for the derivation of model equations governing the dynamics of a phase-separating interface in the following. The succeeding chapter will make use of it to develop the formalism by which the potential is linked to the dynamics of the system in detail. Essentially, the above concepts have to be extended to the theory of irreversible thermodynamic, which opens a way to derive the relevant transport equations.

---

[6] In the context of solidification *latent heat* refers to thermal energy which is released at the interface due to the phase change. Thus a transformation of the inner energy of the system naturally involves a transformation of the latent heat, as well.
[7] Note that within this context the dimensionless latent heat is normalized to 1.
[8] by proceeding just as in (3.8) and (3.9)

4. Thermodynamic Concepts of Phase-Field Modeling

Thermodynamics provides functionals like the Gibb's free energy discussed in detail in the previous chapter. They behave like Lyapunov functions changing monotonically in time. In that sense, thermodynamics provides the basis for a variational approach. If an extremal principle is applied to the rate at which some total (free) energy decreases, then this yields the dynamic equations governing the evolution of the physical system under consideration. The basic theory of how to derive transport equations for irreversible thermodynamic transport processes dates back to Onsager in 1931 [227]. Subsequently it was refined and extended by different other authors [91, 140, 237]. The theory is based on the assumption that the system under consideration is close to an equilibrium condition, so that *locally* thermodynamic equilibrium can be assumed. As a consequence the concepts of equilibrium thermodynamics can be extended to these non-equilibrium conditions based on linearizations.

However, the Onsager theory is certainly only one step in the direction of a diffuse interface model for a moving, phase-separating surface. It gives an idea of how transport equations can be obtained from the thermodynamic functions of a system. On the other hand it does not contain any notion of a phase-field variable as introduced in Chap. 1. Moreover, kinetic principles as an essential component of the physics of a moving interface are not part of the theory.

Kinetic effects are more likely to be found in phenomenological microstructures evolution theories, which start with constitutive relations and empirical data providing equations and principles for the dynamics. These equations and principles, on the other hand, do not necessarily ensure thermodynamic consistency. Guaranteeing compatibility with the laws of thermodynamics yields additional restrictions, for instance on the elastic coefficients and chemical rate constants. In this sense, thermodynamic laws alone are not sufficient to derive the dynamics of microstructural evolution. Nevertheless any comprehensive approach to model interfacial growth problems has to be compatible with them.

This raises the question: What aspects does a diffuse interface approach to moving boundary problems have to fulfill to qualify as comprehensive approach for such problems? And: How does a thermodynamic foundation help with respect to that point?

To address these questions the first section of this chapter will be devoted to the derivation of transport equations from an underlying thermodynamic potential based on Onsager's theory. In the second section this approach is extended to account for a phase-field variable. The third section finally deals with the question of thermodynamic consistency.

4.1 Derivation of Transport Equations

4.1.1 Considering Conserved Quantities Only

The starting point of the concepts developed in this section is the knowledge of a thermodynamic potential of the system under consideration, e.g. the Gibbs free energy or the entropy. I will refer to it as

$$\mathcal{P}(X_1, ..., X_n) = \int_V P(X_1, ..., X_n) \mathrm{d}V \ . \tag{4.1}$$

Here P is the respective density function. $X_1, ..., X_n$ constitute the set of relevant extensive variables normalized to a unit volume. Within this section I assume X_i to be *locally conserved*. Because of the identity

$$\mathrm{d}P(X_1, ..., X_n) = \sum_{i=1}^{n} \frac{\partial P}{\partial X_i} \mathrm{d}X_i \equiv \sum_{i=1}^{n} F_i \mathrm{d}X_i \tag{4.2}$$

the total density flow $\mathbf{J}_P$ of P results from the density flows $\mathbf{J}_i$ of X_i as

$$\mathbf{J}_P = \sum_{i=1}^{n} F_i \mathbf{J}_i \ . \tag{4.3}$$

Here the conservation equations which have to be fulfilled by the extensive variables can be written as

$$\frac{\partial X_i}{\partial t} + \nabla \cdot \mathbf{J}_i = 0 \qquad (i = 1, ..., n) \ . \tag{4.4}$$

Therefore the rate of the total production $\frac{\mathrm{d}P}{\mathrm{d}t}$ per volume element obeys the equation

$$\dot{P} \equiv \frac{\mathrm{d}P}{\mathrm{d}t} \quad = \quad \frac{\partial P}{\partial t} + \nabla \cdot \mathbf{J}_P \tag{4.5}$$

$$\stackrel{(4.2),(4.3)}{=} \sum_{i=1}^{n} \left(F_i \frac{\partial X_i}{\partial t} + (\nabla F_i) \mathbf{J}_i + F_i \nabla \cdot \mathbf{J}_i \right) \tag{4.6}$$

$$\stackrel{(4.4)}{=} \sum_{i=1}^{n} (\nabla F_i) \mathbf{J}_i \ . \tag{4.7}$$

$((\nabla F_i)$ are called affinities of the potential.) Now assuming that the system under consideration is a Markovian system and that the flows $\mathbf{J}_k$ depend only on the *actual* values of the F_i as well as on the affinities, one can write

$$\mathbf{J}_k = \mathbf{J}_k(\nabla F_1, ..., \nabla F_n, F_1, ..., F_n) \ . \tag{4.8}$$

Since we are close to thermodynamic equilibrium it is justified to expand $\mathbf{J}_k$ around its equilibrium value $\mathbf{J}_k = 0$ and to neglect second and higher order terms. The assumption that vanishing affinities also imply vanishing $\mathbf{J}_k$ yields a gradient expansion

$$\mathbf{J}_k = -\sum_{i=1}^{n} L_{ik} \nabla F_i \tag{4.9}$$

with

$$L_{ik} \equiv -\frac{\partial \mathbf{J}_k}{\partial (\nabla F_i)}\Big|_{\mathbf{J}_k=0} \ . \tag{4.10}$$

L_{ik} are called Onsager coefficients. Within the context of the following they obey the Onsager reciprocity relations [227]

$$L_{ik} = L_{ki} \ . \tag{4.11}$$

Inserting relation (4.7) into continuity equation (4.4) yields the desired transport equations

$$\frac{\partial X_k}{\partial t} = \sum_{i=1}^{n} \nabla \cdot (L_{ik} \nabla F_i) \ . \tag{4.12}$$

A second way to obtain this transport equation is based on the principle of minimal energy dissipation postulated by Onsager as well [227]:

$$\int_V \Phi \mathrm{d}V \overset{!}{=} \text{extremum} \ , \tag{4.13}$$

where Φ is the so-called dissipation function. Within the context of evolution of conserved extensive quantities Φ is defined as

$$2\Phi \equiv -\sum_{i,k=1}^{n} L_{ik}(\nabla F_i)(\nabla F_j) \overset{.}{=} \dot{P} \ . \tag{4.14}$$

When applying the variational principle one has to take into account, that the interest is not to obtain a free extremum of (4.13). Rather (4.13) has to be solved together with the condition

$$\dot{P} = 2\Phi \ , \tag{4.15}$$

since only expressions satisfying the linear thermodynamic relations proposed by Onsager qualify as solutions. Introducing the Lagrangian parameter λ one obtains

$$\delta \int_V \Phi + \lambda(2\Phi - \dot{P})dV = 0 \tag{4.16}$$

or

$$\int_V \left(\frac{\partial \Phi}{\partial(\nabla F_i)} + \lambda \left(2\frac{\partial \Phi}{\partial(\nabla F_i)} - \frac{\partial \dot{P}}{\partial(\nabla F_i)} \right) \right) \delta(\nabla F_i)dV = 0 \ , \tag{4.17}$$

respectively. Since the latter equation has to be fulfilled for all virtual displacements $\delta(\nabla F_i)$, (4.7) and (4.14) can be employed to formulate

$$-(2\lambda+1) \sum_{i,k=1}^{n} L_{ik}(\nabla F_i)(\nabla F_k) - \lambda \sum_{i=1}^{n} \nabla F_i \mathbf{J}_i = 2(2\lambda+1)\Phi - \lambda\dot{P} = 0 \ . \tag{4.18}$$

Comparing this with (4.15) yields $\lambda = -1$. Finally inserting into (4.16) results in the free variation equation

$$\delta \int_V (\dot{P} - \Phi)dV = 0 \ . \tag{4.19}$$

Now (4.7) and (4.14) together yield

$$0 = \delta \int_V \sum_{i=1}^{n} \mathbf{J}_i \nabla F_i + \frac{1}{2} \sum_{i,k=1}^{n} L_{ik}(\nabla F_i)(\nabla F_k)dV \tag{4.20}$$

$$= -\delta \int_V \left\{ \sum_{i=1}^{n} (\nabla \mathbf{J}_i)F_i - \frac{1}{2} \sum_{i,k=1}^{n} L_{ik}(\nabla F_i)(\nabla F_k) \right\} dV$$

$$+\delta \oint_{\partial V} \sum_{i=1}^{n} \mathbf{J}_i F_i d\mathbf{S} \ . \tag{4.21}$$

The variation over the surface integral vanishes, since the boundaries of integration are fixed. Together with continuity equation (4.4) this yields

$$\delta \int_V \left\{ \sum_{i=1}^{n} \frac{\partial X_i}{\partial t} F_i + \frac{1}{2} \sum_{i,k=1}^{n} L_{ik}(\nabla F_i)(\nabla F_k) \right\} dV = 0 \ , \tag{4.22}$$

with Lagrangian density

$$L = \frac{\partial X_i}{\partial t} F_i + \frac{1}{2} \sum_{i,k=1}^{n} L_{ik}(\nabla F_i)(\nabla F_k).$$

To fulfill (4.22), the Euler–Lagrange equation given by

$$\frac{\partial L}{\partial F_k} - \frac{\mathrm{d}}{\mathrm{d}x} \frac{\partial L}{\partial \frac{\partial F_k}{\partial x}} - \frac{\mathrm{d}}{\mathrm{d}y} \frac{\partial L}{\partial \frac{\partial F_k}{\partial y}} - \frac{\mathrm{d}}{\mathrm{d}z} \frac{\partial L}{\partial \frac{\partial F_k}{\partial z}} = 0 \tag{4.23}$$

has to be satisfied. Here x, y and z are the relevant space variables. The variation has to be carried out solely with respect to the F_i. The gradient terms, which denote the driving forces of the system, are not taken into account. For this case the Euler–Lagrange equation can be rewritten as

$$\frac{\partial L}{\partial F_k} - \frac{\partial}{\partial x}\frac{\partial L}{\partial \frac{\partial F_k}{\partial x}} - \frac{\partial}{\partial y}\frac{\partial L}{\partial \frac{\partial F_k}{\partial y}} - \frac{\partial}{\partial z}\frac{\partial L}{\partial \frac{\partial F_k}{\partial z}} = 0 . \tag{4.24}$$

This results directly in the desired transport equations:

$$\frac{\partial X_k}{\partial t} = \sum_{i=1}^{n} \nabla(L_{ik}\nabla F_i) . \tag{4.25}$$

Obviously (4.25) is identical to (4.12), thus revealing the equivalence of the two approaches.

4.1.2 Extension to Non-conserved Quantities

If the density function P of the last section does not only contain conserved quantities $\{X_1, ..., X_m\}$, but also non-conserved quantities $\{Y_{m+1}, ..., Y_n\}$, the calculations of the previous section have to be extended. In this case one can write

$$P = P(X_1, ..., X_m, Y_{m+1}, ..., Y_n) . \tag{4.26}$$

In direct analogy to the procedure of the previous section the extended forms of (4.2), (4.8) and (4.9) can be obtained as

$$dP(X_1, ..., X_m, Y_{m+1}, ..., Y_n) = \sum_{i=1}^{m} \frac{\partial P}{\partial X_i}dX_i + \sum_{j=m+1}^{n} \frac{\partial P}{\partial Y_i}dY_i \tag{4.27}$$

$$\equiv \sum_{i=1}^{m} F_i dX_i + \sum_{j=m+1}^{n} G_j dY_j \tag{4.28}$$

$$\mathbf{J}_k = \mathbf{J}_k(\nabla F_1, ..., \nabla F_m, \nabla G_{m+1}, ..., \nabla G_n, F_1, ..., F_m, G_{m+1}, ..., G_n)\,(4.29)$$

$$\mathbf{J}_k = -\sum_{i=1}^{m} L_{ik}\nabla F_i - \sum_{j=m+1}^{n} L_{jk}\nabla G_j . \tag{4.30}$$

In (4.29) and (4.30) flow functions $\mathbf{J}_k$ can only be formulated for conserved quantities, thus index k runs from 1 to m. Inserting the above relations into the continuity equation results in the transport equations for the *conserved* quantities of the system

$$\frac{\partial X_k}{\partial t} = \sum_{i=1}^{m} \nabla(L_{ik}\nabla F_i) + \sum_{j=m+1}^{n} \nabla(L_{jk}\nabla G_j) \qquad (k = 1, ..., m) . \tag{4.31}$$

We are left with looking for transport equations for the *non-conserved* quantities. Generally temporal changes of the Y_j depend on all of the F_i and G_j.

Since we assume that deviation from thermodynamic equilibrium is small, it is sufficient to consider only non-vanishing terms of leading-order in F_i and G_i. At thermodynamic equilibrium $\mathcal{P}$ takes an extremal value, so that a necessary condition reads

$$\frac{\delta \mathcal{P}}{\delta X_i} = \frac{\delta \mathcal{P}}{\delta Y_j} = 0 \,. \tag{4.32}$$

(The extremal principle underlying this formulation and its relation to the Euler–Lagrange equation is elucidated further in Appendix A of this chapter.) If these conditions are not fulfilled for all Y_j, usually the system has to be considered as driven out of equilibrium. As a consequence the Y_j will display variations in time. Therefore, within the framework of linear approximation, we can assume

$$\frac{\partial Y_k}{\partial t} \propto -\frac{\delta \mathcal{P}}{\delta Y_j} \qquad (j = m+1, ..., n) \tag{4.33}$$

for all $k \in \{m+1, ..., n\}$.

If the equilibrium state is perturbed by deviations of one conserved quantity it is not sufficient to change $\frac{\delta \mathcal{P}}{\delta X_i}$ (as one can take from the continuity equation). Rather the gradients of this quantity have to be varied as well. To ensure spatial isotropy of temporal changes in the non-conserved quantities, to leading-order the dependence on the conserved quantity has to satisfy

$$\frac{\partial Y_k}{\partial t} \propto \nabla^2 \frac{\delta \mathcal{P}}{\delta X_i} \qquad (i = 1, ..., m) \,. \tag{4.34}$$

Together (4.31), (4.33) and (4.34) result in

$$\frac{\partial X_k}{\partial t} = \sum_{i=1}^{m} \nabla(L_{ik} \nabla \frac{\delta \mathcal{P}}{\delta X_i}) + \sum_{j=m+1}^{n} \nabla(L_{jk} \nabla \frac{\delta \mathcal{P}}{\delta Y_j}) \qquad (k = 1, ..., m)$$

$$\frac{\partial Y_k}{\partial t} = \sum_{i=1}^{m} \nabla(\kappa_{ik} \nabla \frac{\delta \mathcal{P}}{\delta X_i}) - \sum_{j=m+1}^{n} (\kappa_{jk} \frac{\delta \mathcal{P}}{\delta Y_j}) \qquad (k = m+1, ..., n) \,.$$

$$\tag{4.35}$$

Here the non-diagonal elements L_{ij} and κ_{ij} ($i \neq j$) describe cross coupling effects. In the framework of solidification these are known to result in the Soret [177] or the Dufour [152] effect.

If one assumes the system to depend on one non-conserved quantity Φ and two conserved quantities C and Q only, as well as the energy functional to depend explicitly on the space–dependent observables and their gradients only, the impact of the cross coupling effects can be analyzed starting from the quadratic approximation

$$P = \frac{A_1}{2}\Phi^2 + \frac{A_2}{2}(\nabla\Phi)^2 + \frac{A_3}{2}C^2 + \frac{A_4}{2}(\nabla C)^2 + \frac{A_5}{2}Q^2 + \frac{A_6}{2}(\nabla Q)^2. \quad (4.36)$$

Again a close-to-equilibrium condition is assumed. The dynamics resulting from this reads as follows

$$\frac{\partial}{\partial t}\begin{pmatrix} \Phi \\ C \\ Q \end{pmatrix} = \quad (4.37)$$

$$\begin{pmatrix} \kappa_{11}(A_2\nabla^2 - A_1) & \kappa_{12}\nabla^2(A_3 - A_4\nabla^2) & \kappa_{13}\nabla^2(A_5 - A_6\nabla^2) \\ -L_{21}\nabla^2(A_1 - A_2\nabla^2) & L_{22}\nabla^2(A_3 - A_4\nabla^2) & L_{23}\nabla^2(A_5 - A_6\nabla^2) \\ -L_{31}\nabla^2(A_1 - A_2\nabla^2) & L_{32}\nabla^2(A_3 - A_4\nabla^2) & L_{33}\nabla^2(A_5 - A_6\nabla^2) \end{pmatrix}\begin{pmatrix} \Phi \\ C \\ Q \end{pmatrix} .$$

Employing the ansatz

$$\begin{pmatrix} \Phi \\ C \\ Q \end{pmatrix} = \begin{pmatrix} \Phi_0 \\ C_0 \\ Q_0 \end{pmatrix} e^{i\mathbf{k}\mathbf{r}+\omega t} \quad (4.38)$$

a periodic perturbation of the equilibrium situation with growth rates in terms of the wave vector $\mathbf{k}$ can be carried out. Assuming constant phenomenological coefficients, (4.37) and (4.38) result in

$$\det \begin{vmatrix} \omega + \kappa_{11}(A_1 + A_2 k^2) & \kappa_{12}(A_3 k^2 + A_4 k^4) & \kappa_{13}(A_5 k^2 + A_6 k^4) \\ L_{21}(A_1 k^2 + A_2 k^4) & \omega + L_{22}(A_3 k^2 + A_4 k^4) & L_{23}(A_5 k^2 + A_6 k^4) \\ L_{31}(A_1 k^2 + A_2 k^4) & L_{32}(A_3 k^2 + A_4 k^4) & \omega + L_{33}(A_5 k^2 + A_6 k^4) \end{vmatrix} = 0$$
$$(4.39)$$

with $k = |\mathbf{k}|$. If $\mathbf{k}$ represents a wave length which is large compared to the system's inherent length scales[1], terms of fourth order in k and higher can be neglected. Thus the above determinant yields:

$$0 = \omega^3 + (\kappa_{11}(A_1 + A_2 k^2) + L_{22}A_3 k^2 + L_{33}A_5 k^2)\omega^2$$
$$+(L_{22}A_3 + L_{33}A_5)\kappa_{11}A_1 k^2\omega ,$$

with three solutions

$$\omega = \begin{Bmatrix} 0 \\ -\kappa_{11}(A_1 + A_2 k^2) \\ -(L_{22}A_3 + L_{33}A_5)k^2 \end{Bmatrix} + O(k^4) . \quad (4.40)$$

---
[1] as e.g. the capillary length (i.e., $k \ll 1$)

According to this result, for $k \ll 1$ the growth rate ω depends only on the potential parameters A_i as well as the phenomenological coefficients κ_{11}, L_{22} and L_{33}, thus the diagonal elements. The dependence on non-diagonal elements is of higher order. Therefore it is justified to neglect the cross coupling terms in (4.35).

Moreover one can deduce from the fact that (4.40) is independent of A_4 and A_6 that the dependence of the energy functional on the derivative of the *conserved quantities* can be neglected as well. Thus the dynamics of the locally conserved observables in (4.37) is reduced to a diffusion equation[2]. Defining

$$L_k \equiv L_{kk} \qquad ; \qquad \kappa_k = \kappa_{kk} \qquad (4.41)$$

(4.35) can now be simplified to read

$$
\begin{aligned}
\frac{\partial X_k}{\partial t} &= \nabla(L_k \nabla \frac{\delta \mathcal{P}}{\delta X_k}) \quad (k = 1, ..., m) \\
\frac{\partial Y_k}{\partial t} &= -(\kappa_k \frac{\delta \mathcal{P}}{\delta Y_k}) \qquad (k = m+1, ..., n) \ .
\end{aligned}
\qquad (4.42)
$$

Here the L_k are – apart from a term resulting from the derivative of the variation – identical with the diffusion constants D_k.

The diffusion constants D_k usually depend on the X_k, too. E.g., for mass transport in dilute binary mixtures the diffusion constant is proportional to the impurity concentration C. This can be obtained from setting the flow density proportional to the product of a mobility assumed to be approximately constant and the concentration [185, 210, 295]. Within this limit the thermodynamic potential contains a term proportional to $C \ln C$. As a consequence the dynamics of the concentration field is that of a diffusion equation with an effective constant diffusion coefficient.

Equations (4.35) and (4.42) – depending on whether one intends to include cross coupling effects or not – contain the basic variational principles underlying the concept of diffuse interface modeling developed in the following.

4.2 Introducing the Phase-Field Variable $\varPhi$

The previous section gave a description of how transport equations of a thermodynamic quantity can be obtained by the variation of an appropriate thermodynamic potential with respect to that variable. The introductory chapter,

---

[2] If one is explicitly interested in a study of cross coupling effects, terms of order k^4 have to be taken into account. As a consequence the "diffusion equation" has to be supplemented by ∇^4-terms, as well.

on the other hand, gave a notion of a *phase-field* variable Φ as an essential ingredient of diffuse interface models. Thus if variational principles are to be employed to obtain diffuse interface models, the thermodynamic potential has to be a function of the *phase-field* variable, too. Accordingly one would formulate an ansatz:

$$\mathcal{P}(X_1, ..., X_m, Y_{m+1}, ..., Y_n, \Phi) = \int_V P(X_1, ..., X_m, Y_{m+1}, ..., Y_n, \Phi) \mathrm{d}V .$$

$$(4.43)$$

The inclusion of Φ into a thermodynamic potential as in (4.43) refers back to work done by Fix [119], by Langer [200] and by Collins and Levine [74] in the eighties. The basic idea seems to have been introduced by Landau and Khalatnikov [271] in 1954 to formulate their theory for the absorption of sound in liquid helium. Since then it has been used and developed further by several authors (e.g. [33, 124, 263, 273]) in the context of phase transitions. In this context basic ideas were formulated by picturing the underlying thermodynamic potential $\mathcal{P}$ to be the Gibbs' free energy $\mathcal{G}$ with free energy density G.

Extending the Landau–Khalatnikov ansatz to describe *non-equilibrium* growth phenomena is motivated by the success of introducing such an additional order parameter Φ in the context of the above equilibrium thermodynamic phenomena[3]. In the sense that for non-equilibrium systems the Φ-dependence of the Gibbs free energy cannot be obtained from microscopic first principle considerations, this introduction of Φ remains phenomenological. Certainly this raises the question how to verify models based on a potential as given in (4.43) in a non-equilibrium system. Different approaches to handle the question of verification have been pointed out in the introductory chapter of this work. Here I will describe the basic ideas of how to construct an appropriate Φ-dependence of the energy functional based on work done by Cahn and Hilliard [54, 55] and Allen and Cahn [7]. Further-reaching verification issues will be treated in succeeding sections.

In its generic form the Cahn–Hilliard equation reads

$$\frac{\partial C}{\partial t} = M_C \left(\nabla^2 f'(C) - \xi_C \nabla^4 C \right) .$$

$$(4.44)$$

As such it describes the evolution of a *conserved* concentration field C during phase separation. M_C has to be understood as a constant mobility, f is a free energy. The term $\xi_C \nabla^4 C$ arises as a "penalty" term from the extra energetic cost associated with the transition region between the two phases.

---

[3] Note that to derive the transport equations of a thermodynamic field in the context of a moving boundary system, there is an underlying assumption of instantaneous relaxation of fields to the respective interface position. In that sense the assumption of being close to *local* thermodynamic equilibrium as necessary for the calculations of the previous section, still holds for those fields. It is the moving interface itself which turns the system into a system which is driven out of equilibrium *globally*.

The Cahn–Hilliard equation applies for example to the phenomenon of spinodal decomposition. This is discussed in more detail in Appendix B of this chapter.

The Allen–Cahn equation, on the other hand, describes the evolution of a *non-conserved* order field during anti-phase domain coarsening. Amongst others things it has been used as a model equation for grain growth. A detailed description of this example is given in Appendix C. In its generic form the Allen–Cahn equation can be written as

$$\frac{\partial \eta}{\partial t} = M_\eta \left(\xi_\eta \nabla^2 \eta - f'(\eta) \right) . \tag{4.45}$$

Here η denotes a local order parameter. It can be identified with the phase-field variable Φ appearing in the context of diffuse interface modeling. One would picture the latter to vary smoothly from a fixed value in one phase to another fixed value in the other phase as pictured in Fig. 4.1. [4]

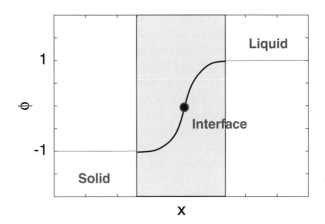

Fig. 4.1. In the context of diffuse interface modeling the phase variable Φ varies smoothly from a fixed value in one phase to another fixed value in the other phase. Here $\Phi = -1$ is chosen to represent the solid phase, $\Phi = 1$ the liquid phase.

The Cahn–Hilliard equation (4.44) and the Allen–Cahn (4.45) equation differ from each other by the fact that the former satisfies an additional conservation condition. The diffuse interface models derived from a Cahn–Hilliard equation can be generalized to the viscous Cahn–Hilliard equations.

---

[4] I refrain from calling Φ itself an order parameter to point out the difference between equilibrium phase transitions and the conditions of interfacial growth, i.e. a non-equilibrium conditions in which the introduction of Φ has to be motivated slightly differently than in the context of the former. This is discussed in some more detail in the remainder of this section.

These contain the Cahn–Hilliard equation and the Allen–Cahn equation in respective limits. In that sense both equations can be treated equally on the basis of one generalized diffuse interface approach. The basic feature of such a generalized description assumes that G is controlled by one contribution originating from a potential $V(\Phi)$ which has two local minima at Φ_L and Φ_S. Φ_L and Φ_S refer to the two separated phases, here liquid and solid. They are also the two homogeneous state solutions the system can take. In the following I will assume Φ_S to be -1 and Φ_L to be 1. $V(\Phi)$ is given by a double–well potential as depicted in Fig. 4.2.

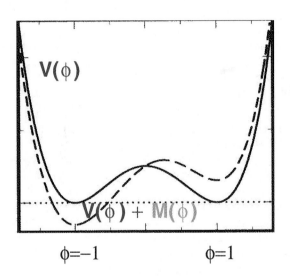

Fig. 4.2. The solid line visualizes the double–well potential $V(\Phi)$ with the two minima at $\Phi = -1$ and $\Phi = 1$ representing the homogeneous states Φ_S and Φ_L, respectively. The dashed line, on the other hand, displays how this potential is unsymmetrized to favor one of the two phases and thereby trigger its growth at the cost of the other. This effect is due to the contribution from $M(\Phi)$.

Assuming that Φ is an analytical function of space and time, $V(\Phi)$ can be expanded into a Taylor series. Such an expansion up to fourth order in Φ yields a double–well potential. Terms of uneven exponent vanish, since $V(\Phi)$ represents a symmetric state, in which the two different phases are energetically equal. As a consequence $V(\Phi)$ has to be invariant with respect to the transformation $\Phi \to -\Phi$. If the two minima of $V(\Phi)$ define the energy level 0, then $V(\Phi)$ reads

$$V(\Phi) = \frac{V_0}{4}(\Phi^2 - 1)^2 \,, \tag{4.46}$$

with a scale factor V_0. Dropping the assumption that $V(\Phi)$ originates from a Taylor expansion, and retaining only the requirement of two minima at ± 1 [23], one can define $V(\Phi)$ section-wise as a construct of two parabolas [37, 38]. In that way it is possible to avoid a Φ^3 inhomogeneity in the derived model equations, which renders analytical treatment more difficult. For numerical simulations, on the other hand, the instability at the crossing point of the two parabolas is undesired, so that for those one usually uses a form of $V(\Phi)$ as given by (4.46).

To be able to consider *inhomogeneous* states as well, the energy density functional has to depend also on the gradient of the phase-field variable. Lowest order terms which are compatible with invariance under rotation and invariance under translation are $\nabla^2\Phi$ and $(\nabla\Phi)^2$. For the calculation of $\mathcal{G}$ the relevant volume integrals of both terms proof to be identical except for a surface integral. This surface integral, however, describes only boundary effects. It vanishes if the phase boundary is completely inside of the integration volume and thus can be neglected. Therefore it is sufficient to simply extend (4.46) by $(\nabla\Phi)^2$ and express G as:

$$G(\Phi, \mathbf{X}, \mathbf{r}) = \frac{\xi^2}{2}(\nabla\Phi)^2 + V(\Phi) , \qquad (4.47)$$

where I have summarized the variables $X_1, ..., X_m$ and $Y_{m+1}, ..., Y_n$ of (4.43) as vector $\mathbf{X}$ and introduced the vector $\mathbf{r}$ to denote space coordinates.

The potential (4.47) still refers to an equilibrium condition. It is possible to give a microscopic derivation for this potential as well. It can be obtained on the basis of a lattice model for a two-component equilibrium thermodynamic system [48, 93, 118, 154]. Within such a model each of the lattice squares $\mathbf{x}$ of a rectangular grid is occupied by a variable $\phi(\mathbf{x}) \in \{-1; 1\}$. The variable ϕ distinguishes between the two possible states. For the energy of a given configuration the following ansatz is chosen

$$G = -\frac{1}{2}\sum_{\mathbf{x},\mathbf{x}'} J(\mathbf{x} - \mathbf{x}')\phi(\mathbf{x})\phi(\mathbf{x}') - H\sum_{\mathbf{x}} \phi(\mathbf{x}) . \qquad (4.48)$$

The first term denotes the interaction energy originating from the coupling term $J(\mathbf{x} - \mathbf{x}')$, the second term contains the energy contribution from external fields.

Often it is feasible to treat ϕ as a continuous variable. In this case it is necessary to include a function $W(\phi(\mathbf{x}))$. W is a weight function which determines the physically favored values of ϕ. With the introduction of W one obtains the Landau–Ginzburg–Wilson functional as

$$G = -\frac{1}{2}\sum_{\mathbf{x},\mathbf{x}'} J(\mathbf{x} - \mathbf{x}')\phi(\mathbf{x})\phi(\mathbf{x}') - H\sum_{\mathbf{x}} \phi(\mathbf{x}) + \sum_{\mathbf{x}} W(\phi(\mathbf{x})) . \qquad (4.49)$$

For $W(\phi(\mathbf{x}))$ a symmetric potential with second- and fourth-order terms is the suitable form to describe a two-state system. For the case of a short-range interaction $J(\mathbf{x} - \mathbf{x}')$ (e.g. nearest neighbor interaction) as well as vanishing external fields, (4.49) can be transformed into a continuity equation analogous to (4.47) ($\phi \to \Phi$) [48, 154]. Thus for *thermodynamic equilibrium* configurations microscopic and macroscopic derivation yield the same ansatz for $G(\Phi, \mathbf{X}, \mathbf{r})$. Furthermore the microscopic approach provides additional insight into the nature of diffuse interface models. It yields an understanding, that the finite interface thickness of a diffuse interface model and the surface energy term originate from finite correlation lengths on a microscopic scale. Moreover – assuming nearest neighbor interaction and constant coupling constant J – interface thickness and surface energy are proportional to J.

For non-equilibrium conditions a coherent microscopic derivation is missing. From a macroscopic point of view, (4.47) can be extended to account for non-equilibrium conditions by introducing a further energetic term $M(\Phi, \mathbf{X})$ such that

$$G(\Phi, \mathbf{X}, \mathbf{r}) = \frac{\xi^2}{2}(\nabla \Phi)^2 + V(\Phi) + M(\Phi, \mathbf{X}) \ . \tag{4.50}$$

The contribution $M(\Phi, \mathbf{X})$ depends on the special physical properties of the system under consideration, e.g. the underlying phase diagram. Together with $V(\Phi)$ this additional term M results in a double–well potential, which is no longer symmetrical with respect to $\Phi \to -\Phi$. This is depicted in Fig. 4.2. The dynamic equations obtained from (4.50) owe the driving force, which moves the interface, to the term $M(\Phi, \mathbf{X})$. In this context the evolutionary equation for the phase-field variable Φ can be interpreted as a mean-field approximation of the non-equilibrium interface dynamic.

Analytical as well as numerical treatment of model equations obtained from (4.50) is simplified by the localizing of the minima of the unsymmetric double–well potential $V(\Phi) + M(\Phi, \mathbf{X})$ at ± 1. This yields

$$\frac{\partial M(\Phi, \mathbf{X})}{\partial \Phi} \Big|_{\Phi = \pm 1} = 0 \ . \tag{4.51}$$

Equation (4.51) has to be valid for arbitrary $\mathbf{X}$. One can ensure the latter choosing the ansatz

$$M(\Phi, \mathbf{X}) = \sum_{k=1}^{n} M_{\Phi,k}(\Phi) M_{\mathbf{X},k}(\mathbf{X}) \ , \tag{4.52}$$

where

$$\frac{\partial M(\Phi)}{\partial \Phi} \Big|_{\Phi = \pm 1} = 0 \ . \tag{4.53}$$

In this way one avoids the undesired effect that within a pure liquid or a pure solid, respectively, changes to the equilibrium value of the phase field result in emission of latent heat.

If one would classify the dynamics of interfacial growth we are interested in here according to a scheme of dynamic universality classes proposed by

Hohenberg and Halpering in [146] [5] it would correspond to their "model C". A classification as in [146] can be done based on a thermodynamic functional as $\mathcal{G}(\mathbf{X}, \Phi)$ in (4.47). However, the specific functional employed in [146] is the Helmholtz free energy functional, which I will denote by $\mathcal{A}(\mathbf{X}, \Phi)$ in the following. Even without a precise expression for $\mathcal{A}(\mathbf{X}, \Phi)$ one can distinguish between three different types of evolution equations obtained for the phase-field variable Φ in the following manner:

1. **Model A** states a purely relaxational evolution of Φ. It can be written as

$$\frac{\partial \Phi}{\partial t} = -K \frac{\delta \mathcal{A}}{\delta \Phi} \; . \tag{4.54}$$

   Here K is a positive coefficient which might depend on Φ and on the temperature. Equation (4.54) can be understood to model systems where an internal order parameter expressed by Φ characterizes a non-equilibrium structure which eventually[6] approaches an equilibrium value obtained by minimizing the free energy function.

2. **Model B** applies to a system in which an order parameter represents an equilibrium thermodynamic variable, e.g. the concentration. In this case a generic evolution equation based on $\mathcal{A}$ reads

$$\frac{\partial \Phi}{\partial t} = \nabla \cdot (M \nabla \frac{\delta \mathcal{A}}{\delta \Phi}) \; , \tag{4.55}$$

   where M is a positive coefficient related to diffusivity. This expression was introduced by Cahn as a model for spinodal decomposition in 1961 [57].

3. **Model C** is the most specific of the three generic models. It describes the dynamics of a system for which one order parameter as given by Φ is insufficient to describe the local state of the system. Additional fields need to be incorporated into the coupled model equations as well. The most popular model for this case involves the temperature T in addition to Φ [143] and thus applies to solidification of pure undercooled melts:

$$\alpha \xi^2 \frac{\partial \Phi}{\partial t} = \xi^2 \nabla^2 \Phi + g(\Phi) - f(T) \tag{4.56}$$

$$\frac{\partial f}{\partial t} - \lambda \frac{\partial \Phi}{\partial t} = \nabla^2 f \; . \tag{4.57}$$

   Here f is a strictly increasing function of the temperature, g is a negative double–well function, α, ξ and λ are positive constants.

As stated above, the coupled set of equations (4.56) and (4.57) can be viewed as one of the first diffuse interface or *phase-field* models for interfacial growth. Introduced in the same manner as by Halperin *et al.*, it remains a

---

[5] An earlier approach in similar direction can be found in [143].
[6] Here *eventually* refers to the limit $t \to \infty$.

phenomenological approach. The reason is that even though (4.56) is obtained formally by carrying out a variation according to (4.42), (4.57) is constructed merely by common sense introducing the effects due to the moving interface as source term $\lambda\frac{\partial\Phi}{\partial t}$ "by hand". Moreover, strict relaxational behavior of the potential during the course of evolution is not ensured. I will come back to this point in the discussion following (4.62).

In 1990 Penrose and Fife managed to place this model upon a formally more rigorous foundation. To do so they expressed (4.56) and (4.57) in terms of an entropy functional (see [231], p.50). This results in

$$\alpha\xi^2\frac{\partial\Phi}{\partial t} = K\frac{\delta S}{\delta\Phi} \tag{4.58}$$

$$\frac{\partial u}{\partial t} = -\nabla(M\nabla\frac{1}{T}) \ . \tag{4.59}$$

Here K and M are positive constants similar to those above, and u is a homogenous energy density. In [231] (p.51) it is given by

$$u \equiv f(T)\nu(\Phi) + e_\Phi = \frac{\mathrm{d}\frac{a_h(t,\Phi)}{T}}{\mathrm{d}\frac{1}{T}} \ , \tag{4.60}$$

where ν and e_Φ are defined as the number of degrees of freedom per unit volume and the potential energy, respectively. $a_h(t,\Phi)$ is the homogenous contribution to the Helmholtz free energy density. The entropy functional defined by Penrose and Fife (see [231], p.49) reads

$$S = \int_V \{s(u,\Phi) - \frac{1}{2}\tilde{\xi}|\nabla\Phi|^2\}\mathrm{d}V \ . \tag{4.61}$$

Here $\tilde{\xi} = \xi/T$. s represents the homogenous entropy density defined in the usual sense via the thermodynamic relation

$$s = \frac{u}{T} - \frac{a_h(t,\Phi)}{T} \ . \tag{4.62}$$

Thus applying standard thermodynamic relations to "model C", Penrose and Fife used homogeneous densities only.

The great difference between model equations (4.56)–(4.57) compared to (4.58)–(4.59) is that the latter are *thermodynamically consistent*, whereas the former are not. As introduced in Chap. 1 *thermodynamic consistency* refers to a strict relaxation of thermodynamic potentials during the evolution of the respective system.

The fact that models obtained from a Helmholtz free energy (like model equations (4.56) and (4.57)) are not thermodynamically consistent, originates from the point that changes in temperature trigger changes of the inner energy as well. Those changes would have to be integrated into the above derivation. This happens naturally, if one transforms from the Helmholtz free energy as underlying potential to the entropy, as well as from the temperature as independent variable to the inner energy density u.[7]

Remembering that in the beginning of this section the introduction of a phase-field variable Φ in non-equilibrium growth processes could be motivated only phenomenologically, renders the concept of thermodynamic consistency an important backbone of phase-field modeling. It places the usage of a phase field upon a solid theoretical foundation. In this sense diffuse interface models ensuring strict relaxational behavior of the potentials are more but a phenomenological description of an interfacial growth problem.

However, keeping in mind that the classical approach to interfacial growth problems, the so-called *sharp interface* approach (see Chap. 1), is phenomenological in nature, too, employing a diffuse interface approach, it might not always be necessary to formulate the model to be thermodynamically consistent. Just as well it can be justified to work with a phenomenological model as "model C", if the analogy to an established sharp interface model can be proven via an asymptotic matching expansion as described in Chap. 5. From a numerical point of view this is often more efficient. In this sense within the numerical appendix of this book the concept of *universality* of different diffuse interface models, thermodynamically consistent ones as well as phenomenological ones, will be addressed for the very classical phenomenon of purely diffusion limited growth. However, it has to be stressed again that the methodology of thermodynamically consistent diffuse interface modeling is the one, which opens new perspectives if diffuse interface modeling is applied to more complicated physical systems. For such cases many important issues are still researched. Chapter 6 will therefore address the topic of *thermodynamic consistency* again in the context of an example, which demonstrates, how the methodology can be used to learn more about the physics of more complicated interfacial growth phenomena. The remainder of this chapter will merely summarize some approaches to ensure thermodynamic consistency developed so far.

4.3 Thermodynamic Consistency

As described in Sects. 4.1 and 4.2 the starting point to obtain the evolutionary equations of a diffuse interface model is given by the variational principles (4.35) or (4.42), respectively. On the other hand, applying such methods to a somewhat phenomenologically[8] introduced phase-field variable, does not necessarily yield a thermodynamically consistent dynamics. This implies that strict relaxational behavior of the thermodynamic potentials during the course of evolution is not necessarily ensured. The first part of this section

---

[7] just as it was done in (4.58) and (4.59)

[8] The notion *phenomenological* refers to the respective discussions in Sects. 4.1 and 4.2.

describes a comprehensive approach to extend the variational methods of the previous sections in such a way that they guarantee conformance with thermodynamics. This comprehensive approach assumes that interfacial motion is driven either by surface diffusion or by curvature. In that sense this approach is not suitable to model general interfacial growth problems. However, it reveals two important aspects:

1. How kinetics can be integrated into a variational approach, and
2. how thermodynamic consistency can be guaranteed by a variational method.

The approach ensuring the above is called *gradient flow method*. It was developed intensively in the mathematical "moving interface" community in the nineties [8, 63, 265, 268, 269]. An important concept, which arose along with that development is the Cahn–Hoffmann capillary vector to model anisotropic surface properties in the context of curvature driven interfacial motion [58, 59, 145, 267]. It is an essential ingredient of any diffuse interface model for interfacial growth phenomena, which is meant to recover the anisotropy of surface energetics.

As mentioned before, for the inclusion of further physical mechanisms or fields (e.g. a hydrodynamic field) a comprehensive approach is still missing. Nevertheless by expressing local entropy production in terms of the transport quantities of derived evolutionary equations and identifying the forms of these quantities which ensure it to be non-negative, consistent diffuse interface models have been obtained for even more complex systems. One example along that line is the approach by Anderson, McFadden and Wheeler to solidification taking into account hydrodynamic convection [15]. In this case the analysis, in which the thermodynamically consistent diffuse interface model emerges, is case sensitive adapted to the individual physical situation. As an example for such a case sensitive ansatz compared to the above comprehensive framework, the main steps of their model approach is summarized in the second part of this section. The question of what remains to be done to find a general approach for any interfacial growth phenomenon, complex as it might be, is left to Chaps. 6, 7 and 8.

4.3.1 The Gradient Flow Method

The basic idea underlying the gradient flow method is to guarantee consistency with the laws of thermodynamics by constructing a gradient flow which decreases the energy as much as possible at a given time step. This gradient flow is obtained from an inner product which yields a kinetic measure of the time it would take for a system to evolve from one microstructure to another. Mathematically its derivation reads

$$\left.\frac{\partial \mathcal{A}}{\partial t}\right|_{t=0} = \langle \nabla \mathcal{A}, v \rangle . \tag{4.63}$$

Just as in the previous sections $\mathcal{A}$ can be understood to be the functional of a free energy density. For the inner product $\langle \nabla \mathcal{A}, v \rangle$ a norm has to be specified. $\nabla \mathcal{A}$ will depend on the choice of the inner product, i.e.

$$\langle \nabla \mathcal{A}, v \rangle = \langle \nabla_{\langle,\rangle} \mathcal{A}, v \rangle . \tag{4.64}$$

The most important norms for the discussion of interfacial motion are the L^2–norm and the H^{-1}–norm defined by:

$$\langle f, v \rangle_{L^2} = f \cdot v = \int_V fv \mathrm{d}V , \tag{4.65}$$

$$\langle f, v \rangle_{H^{-1}} = \int_V \nabla \phi_f \nabla \phi_v \mathrm{d}V = - \int_V \left[(\nabla^2)^{-1} f \right] v \mathrm{d}V , \tag{4.66}$$

where ϕ_f is a solution of the Poisson equation

$$\nabla^2 \phi_f = f(\mathbf{r}) . \tag{4.67}$$

Applying this formalism to interfacial motion triggered by surface diffusion, weighting of the inner products with a function $M(\mathbf{r})$ allows me to take into account anisotropic mobility constants. This modifies the inner products as

$$\langle f, v \rangle_{L^2, M} = \int_V \frac{fv}{M} \mathrm{d}V , \tag{4.68}$$

$$\langle f, v \rangle_{H^{-1}} = - \int_V \left[(\nabla M \nabla)^{-1} f \right] v \mathrm{d}V . \tag{4.69}$$

In analogy to the discussion following (4.54) of Sect. 4.2 for the free energy density an ansatz $a(\phi(\mathbf{r})) = a_h + \xi^2 \Gamma(\nabla \phi)$ can be chosen. It enters (4.64) via the functional $\mathcal{A} = \int_V a \mathrm{d}V$. Here the dependence of $\xi^2 \Gamma(\nabla \phi)$ on the direction of $\nabla \phi$ is directly proportional to the orientational dependence of the anisotropic interfacial energy density [268]. As in Sect. 4.2 a_h does not need to be specified any further. It is the choice of a specific norm which results in a specific kind of dynamics. An underlying H^{-1}–norm, for example, yields a generalized anisotropic Cahn–Hilliard equation (see (4.44))

$$\frac{\partial C}{\partial t} = \nabla M \nabla \left(f'(C) - \nabla \Gamma \Gamma' \right) . \tag{4.70}$$

On the other hand, applying the L^2–norm for a non-conserved parameter η generalizes the Allen–Cahn equation (see (4.45)) to anisotropic surface physics yielding

$$\frac{\partial \eta}{\partial t} = M_\eta \left(\nabla \Gamma \Gamma' - f'(\eta) \right) . \tag{4.71}$$

Other specifications of $\mathcal{A}$ will result in still different dynamical behavior, e.g. purely relaxational kinetics or the Mullins surface diffusion equation [222]. In any case the dynamics is obtained in a thermodynamically consistent formulation. In this sense and in the sense of inclusion of anisotropic surface physics, even if only applicable to a very restricted set of physical situations, (4.64) results in a more advanced variational approach than the "model C"–type of approach of the previous section.

4.3.2 Entropy Production in Terms of Transport Variables

Since the above gradient flow method does no longer hold, if one turns to more complicated interfacial growth phenomena, one is left with the variational principles of the previous sections and has to ensure thermodynamic consistency additionally by hand. One way to proceed is to follow the formalism of irreversible thermodynamics [91] and embed the variational principles applying to diffuse interface models into this framework. This procedure is for example underlying the work, by which Penrose and Fife managed to put "model C" onto a more fundamental foundation [231] (see Sect. 4.2). It has been introduced to different growth conditions treated on the basis of a diffuse interface approach by several authors since [283, 289]. In a similar manner one might start from transport equations obtained via balance law formulations, express local entropy production in terms of the transport variables and identify the forms, which ensure it to be non-negative. To illustrate this procedure here the basic steps to derive a diffuse interface model for solidification influenced by hydrodynamic convection presented by Anderson *et al.* [15] are summarized in the succeeding paragraphs:

Following [15] I assume that the total entropy $\mathcal{S}$ within a volume $V(t)$ is given by

$$\mathcal{S} = \int_{V(t)} \left[\rho s - \frac{1}{2}\xi_S^2 \Gamma^2(\nabla\Phi) \right] \mathrm{d}V . \tag{4.72}$$

Here ρ and s are density and entropy per unit mass, respectively. The first term in the integrand (i.e. ρs) is the classical entropy density. The second is a non-classical term associated with spatial gradients of the phase field. For simplicity the gradient entropy coefficient ξ_S is assumed to be a constant. Γ is a homogeneous function of degree unity. As introduced in Sect. 4.3.1, this function allows for a general anisotropic surface energy of the solid–liquid interface.

For the total mass $\mathcal{M}$, linear momentum $\mathcal{P}$ and internal energy $\mathcal{E}$ associated with the volume of the material the following ansatz is chosen:

$$\mathcal{M} = \int_{V(t)} \rho \mathrm{d}V , \tag{4.73}$$

$$\mathcal{P} = \int_{V(t)} \rho \mathbf{u} \mathrm{d}V, \tag{4.74}$$

$$\mathcal{E} = \int_{V(t)} \left[\rho e + \frac{1}{2}\rho |\mathbf{u}|^2 + \frac{1}{2}\xi_E^2 \Gamma^2(\nabla\Phi) \right] \mathrm{d}V. \tag{4.75}$$

Here $\mathbf{u}$ is the velocity, e the internal energy density per unit mass and ξ_E the gradient energy coefficient, which is assumed to be constant. The thermodynamic relations

$$\mathrm{d}e = T\mathrm{d}s + \frac{p}{\rho^2}\mathrm{d}\rho + \frac{\partial e}{\partial \Phi}\mathrm{d}\Phi , \tag{4.76}$$

$$e = Ts - p/\rho + \mu \tag{4.77}$$

are assumed to apply locally, where p is the thermodynamic pressure and μ the chemical potential per unit mass.

The physical balance laws for mass, linear momentum and internal energy are given by

$$\frac{d\mathcal{M}}{dt} = 0 \,, \tag{4.78}$$

$$\frac{d\boldsymbol{P}}{dt} = \int_{V(t)} \hat{\mathbf{n}} \cdot \mathbf{m} dV \,, \tag{4.79}$$

$$\frac{d\mathcal{E}}{dt} + \int_{\partial V(t)} \mathbf{q}_E \hat{\mathbf{n}} dA = \int_{\partial V(t)} \hat{\mathbf{n}} \cdot \mathbf{m} \cdot \mathbf{u} dA \,, \tag{4.80}$$

respectively. Here $\hat{\mathbf{n}}$ is the outward unit vector normal to $\partial V(t)$, $\mathbf{m}$ the stress tensor[9], and $\mathbf{q}_E$ the internal energy flux. Momentum balance given by (4.79) requires that the total momentum rate–change of the material's volume results from forces acting on its boundary $\partial V(t)$. External forces such as gravity are neglected. The equation for energy balance (4.80) equates the rate of change of the total internal energy of $V(t)$ plus the energy flux through its boundary to the rate of work of the forces at this boundary $\partial V(t)$.

In addition, entropy balance is given by

$$\frac{d\mathcal{S}}{dt} + \int_{\partial V(t)} \mathbf{q}_S \hat{\mathbf{n}} dA = \int_{V(t)} \dot{s}^{prod} dV(t) \,, \tag{4.81}$$

where $\mathbf{q}_S$ is the entropy flux and $\dot{s}^{prod}$ the local rate of entropy production[10]. The second law of thermodynamics requires it to be non-negative.

To proceed, Anderson, McFadden and Wheeler rewrite the conservation laws (4.78)–(4.81) as differential equations. These are used to express the local entropy production in terms of the fluxes $\mathbf{m}$, $\mathbf{q}_E$, and $\mathbf{q}_S$, as well as the material derivative of $\varPhi$ which is given by $D\varPhi/Dt = \frac{\partial \varPhi}{\partial t} + (\mathbf{u} \cdot \nabla)\varPhi$. Then they identify forms for these quantities which ensure that the local entropy production is non-negative. The fluxes that result from this procedure involve both classical and non-classical contributions. In addition, they obtain an evolution equation for the phase field.

Applying the Reynolds transport theorem [19] to the mass balance law (4.78) yields the continuity equation in its conventional form

$$\frac{D\rho}{Dt} + \rho \nabla \mathbf{u} = 0 \,. \tag{4.82}$$

---

[9] Tensors and vectors are both denoted by bold print. The only exception is the linear momentum $\boldsymbol{P}$, which for clarity purposes is expressed with a vector symbol to distinguish it from the scalar functions $\mathcal{M}$, $\mathcal{E}$ and $\mathcal{S}$ expressed in calligraphic letters as well.

[10] Note that due to the evolving boundary $V(t)$ is changing in time. For simplicity I will drop the dependence on t in the following.

Again the material derivative $\frac{D}{Dt} = \frac{\partial}{\partial t} + (\mathbf{u} \cdot \nabla)$ has been used. Similarly, the linear momentum equation yields the expression

$$\frac{D\mathbf{u}}{Dt} = \nabla \cdot \mathbf{m}, \tag{4.83}$$

where $\nabla \cdot \mathbf{m}$ has components $\partial m_{jk}/\partial x_j$ (m_{jk} denotes the components of $\mathbf{m}$).

The derivation of the energy equation is more involved. It requires to cast (4.80) into

$$\int_V \left[\rho \frac{De}{Dt} + \rho \mathbf{u} \frac{D\mathbf{u}}{Dt} - \nabla \cdot (\mathbf{m} \cdot \mathbf{u}) + \nabla \mathbf{q}_E \right] dV + \frac{d}{dT} \int_V \frac{1}{2} \xi_E^2 \Gamma^2 (\nabla \Phi) dV = 0 . \tag{4.84}$$

Further the identity

$$\frac{d}{dT} \int_V \frac{1}{2} \xi_E^2 \Gamma^2 (\nabla \Phi) dV = \int_V \xi_E Q_G dV , \tag{4.85}$$

which is proven in the appendix of [15], is employed, where

$$Q_G = \nabla \cdot \left(\Gamma \Sigma \frac{D\Phi}{Dt} \right) - \frac{D\Phi}{Dt} \nabla \cdot (\Gamma \Sigma) - \Gamma \nabla \mathbf{u} : \Sigma \otimes \nabla \Phi + \frac{1}{2} \Gamma^2 \nabla \cdot \mathbf{u} . \tag{4.86}$$

Here $\otimes$ refers to the outer tensor product and : to the double contraction of the tensor product. Σ denotes the Cahn–Hoffmann capillary vector mentioned in the introductory part of this section. It follows that

$$\rho \frac{De}{Dt} + \rho \mathbf{u} \frac{D\mathbf{u}}{Dt} - \nabla \cdot (\mathbf{m} \cdot \mathbf{u}) + \nabla \mathbf{q}_E + \xi_E^2 Q_G = 0 . \tag{4.87}$$

Conservation of linear momentum as given by (4.83) can be used to rewrite this equation as

$$\rho \frac{De}{Dt} + \mathbf{u} \cdot (\nabla \cdot \mathbf{m}) - \nabla \cdot (\mathbf{m} \cdot \mathbf{u}) + \nabla \mathbf{q}_E + \xi_E^2 Q_G = 0 . \tag{4.88}$$

With the identity

$$\nabla \cdot (\mathbf{m} \cdot \mathbf{u}) = (\nabla \cdot \mathbf{m}) \cdot \mathbf{u} + \mathbf{m} : \nabla \mathbf{u} \tag{4.89}$$

this simplifies to yield the energy equation as

$$\rho \frac{De}{Dt} + \nabla \cdot \mathbf{q}_E = \mathbf{m} : \nabla \mathbf{u} - \xi_E^2 Q_Q , \tag{4.90}$$

where the double contraction of the tensor product is given by $\mathbf{m} : \nabla \mathbf{u} = m_{kj} \partial u_j / \partial x_k$ (with summation over repeated indices implied). In direct analogy, entropy balance as given by (4.81) results in

$$\rho \frac{Ds}{Dt} + \nabla \cdot \mathbf{q}_S = \dot{s}^{prod} + \xi_S^2 Q_G . \tag{4.91}$$

The thermodynamic relation between Ds/Dt and De/Dt follows from (4.76). It is given by

$$\rho\frac{De}{Dt} = T\frac{Ds}{Dt} + \frac{p}{\rho^2}\frac{D\rho}{Dt} + \frac{\partial e}{\partial \Phi}\frac{D\Phi}{Dt} \ . \tag{4.92}$$

At this point the continuity equations (4.82) and (4.86), as well as (4.90) and (4.92) can be employed to express entropy production as

$$\dot{s}^{prod} = \frac{1}{T}\left\{\mathbf{m} + \xi_F^2\Gamma\Sigma\otimes\nabla\Phi + \left[p - \frac{\xi_F^2}{2}\Gamma^2\right]\mathbf{I}\right\} : \nabla\mathbf{u}$$

$$+\frac{1}{2}\left\{\xi_F^2\nabla(\Gamma\Sigma) - \rho\frac{\partial e}{\partial\Phi}\right\}\frac{D\Phi}{Dt}$$

$$+\nabla\cdot\left(\mathbf{q}_S - \frac{\mathbf{q}_E}{T} - \frac{\xi_F^2}{T}\Gamma\Sigma\frac{D\Phi}{Dt}\right)$$

$$+\left(\mathbf{q}_E + \xi_E^2\Gamma\Sigma\frac{D\Phi}{Dt}\right)\cdot\nabla(\frac{1}{T}) \ . \tag{4.93}$$

Here $2F = 2E + T2S$.

The constitutive equations for the fluxes and $D\Phi/Dt$ can still be specified ensuring positive $\dot{s}^{prod}$

$$\mathbf{m} = \left[p - \frac{\xi_F^2}{2}\Gamma^2\right]\mathbf{I} - \xi_F^2\Gamma\Sigma\otimes\nabla\Phi + \tau \ , \tag{4.94}$$

$$M\frac{D\Phi}{Dt} = \xi_F^2\nabla(\Gamma\Sigma) - \rho\frac{\partial e}{\partial\Phi} \ , \tag{4.95}$$

$$\mathbf{q}_E = \tilde{k}\nabla\frac{1}{T} - \xi_E^2\Gamma\Sigma\frac{D\Phi}{Dt} \ , \tag{4.96}$$

$$\mathbf{q}_S = \tilde{k}\nabla\frac{1}{T} - \xi_S^2\Gamma\Sigma\frac{D\Phi}{Dt} \ . \tag{4.97}$$

Here τ is the viscous stress tensor, which for a Newtonian fluid is given by $\tau = \mu(\nabla\mathbf{u} + \nabla\mathbf{u}^T) + \lambda(\nabla\mathbf{u})\mathbf{I}$. μ and λ are coefficients of viscosity, $\mathbf{I}$ is the unit tensor, and M is a constant positive mobility coefficient. Moreover, a constant value for the thermal conductivity k is chosen corresponding $\tilde{k} = T^2k$. The resulting coupled set of equations is given in [15] by

$$\frac{D\rho}{Dt} = -\rho\nabla\mathbf{u} \ , \tag{4.98}$$

$$\rho\frac{D\mathbf{u}}{Dt} = \nabla\cdot\left[\left(-p + \frac{\xi_F^2}{2}\Gamma^2\right)\mathbf{I} - \xi_F^2\Gamma\Sigma\otimes\nabla\Phi + \tau\right] \ , \tag{4.99}$$

$$M\frac{D\Phi}{Dt} = \xi_F^2\nabla\cdot(\Gamma\Sigma) - \rho\frac{\partial e}{\partial\Phi} \ , \tag{4.100}$$

$$\rho\frac{De}{Dt} = \nabla\cdot[k\nabla T] + \xi_E^2\nabla\cdot(\Gamma\Sigma)\frac{D\Phi}{Dt}$$

$$+\left[\left(-p + \frac{\xi_S^2}{2}\Gamma^2\right)\mathbf{I} - T\xi_S^2\Gamma\Sigma\otimes\nabla\Phi + \tau\right] : \nabla\mathbf{u} \ . \tag{4.101}$$

Equations (4.98)–(4.101) constitute a diffuse interface model for hydro-dynamically influenced solidification. It is a thermodynamically consistent model. Here its derivation serves as an example for a second approach to ensure thermodynamic consistency after the discussion of the gradient flow method in Sect. 4.3.1. It is carried out by expressing entropy production in terms of the transport quantities of derived evolutionary equations and identifying the forms of these quantities that ensure that the local entropy production cannot be negative. One can interpret the thermodynamic consistency of (4.98)–(4.101) as a validation of this model. On the other hand, analyzing such thermodynamically consistent model equations based on the method of matched asymptotic expansion (see Chap. 5) might revise classical sharp interface equations and thus contribute to new insight into the respective interfacial growth phenomena as I will demonstrate in Chap. 6.

4.4 Appendix

Appendix A: Excursion to Some Basic Calculus of Variation

To derive appropriate transport equations for non-conserved thermodynamic variables, in Sect. 4.1.2 we started out from the consideration, that directly at thermodynamic equilibrium the potential $\mathcal{P}$ should take an extremal value. $\mathcal{P}$ is defined (see (4.1) and (4.26)) as

$$\mathcal{P}(X_1, ..., X_m, Y_1, ..., Y_n) = \int_V P(X_1, ..., X_m, Y_1, ..., Y_n)\mathrm{d}V , \qquad (4.102)$$

where P is the respective density function, the X_i with $i = 1, ..., m$ conserved variables and the Y_j $(j = 1, ..., n)$ non-conserved variables. Then requiring

$$\mathcal{P}(X_1, ..., X_m, Y_1, ..., Y_n) \doteq \text{extremum} , \qquad (4.103)$$

is formally a classical variational problem. We could rewrite the variation $\frac{\delta \mathcal{P}}{\delta X_i}$ within (4.32) as

$$\frac{\delta \mathcal{P}}{\delta X_i} := \frac{\mathrm{d}}{\mathrm{d}s} P(X_i + s \cdot \eta)\Big|_{s=0} , \qquad (4.104)$$

where the variable η is defined on V (just as the X_i and Y_j). If X_i minimizes (maximizes) $\mathcal{P}$, then $\frac{\delta \mathcal{P}}{\delta X_i} = 0$ for all such η. The classical strategy for solving the problem

$$\mathcal{P} \overset{!}{=} \text{minimum} \qquad (4.105)$$

consists in solving the respective Euler–Lagrange equations and then investigating whether a solution of the equations is a minimum of $\mathcal{P}$ or not. Assuming that the X_i depend solely on the time t, the corresponding Euler–Lagrange equations read:

$$\frac{\mathrm{d}}{\mathrm{d}t} P_p(t, X_i(t), X_i'(t)) - P_n(t, X_i(t), X_i'(t)) = 0 , \qquad (4.106)$$

where P_n is a vector of the partial derivatives of P with respect to the X_i and P_p an analogous vector of the partial derivatives of P with respect to the X_i', which are the derivatives of the X_i with respect to time. For a derivation of the Euler–Lagrange equations the reader is referred to introductory courses into classical mechanics as e.g. [134]. Here I will conclude this excursion on variations with an example to elucidate the relation between an extremal principle and the Euler–Lagrange equations somewhat further.

Example: The question I want to answer is how to minimize the arc length of the graph of a function $u : [a, b] \to \mathbb{R}$, i.e. the length of a curve $(t, u(t))$, which is a subset of $\mathbb{R}^2$, among all graphs with fixed boundary values $u(a)$, $u(b)$. This leads to the variational problem

$$\int_a^b \sqrt{1 + u'(t)^2} \overset{!}{=} \text{minimum} . \qquad (4.107)$$

To compute the relevant Euler–Lagrange equation we first have to evaluate the partial derivative with respect to u as well as the partial derivative with respect to u'. They read $P_u = 0$ and $P_p = \dfrac{u'(t)}{\sqrt{1+u'(t)^2}}$, respectively. Thus the Euler-Lagrange equation is given by

$$0 = \frac{\mathrm{d}}{\mathrm{d}t} \frac{u'(t)}{\sqrt{1 + u'(t)^2}} = \frac{u''(t)}{\sqrt{1 + u'(t)^2}} - \frac{u'(t)^2 u''(t)}{\left(\sqrt{1 + u'(t)^2}\right)^3} \qquad (4.108)$$

$$= \frac{u''(t)}{\left(\sqrt{1 + u'(t)^2}\right)^3} , \qquad (4.109)$$

i.e.

$$u''(t) = 0 . \qquad (4.110)$$

This implies that u is a straight line between a and b, which in this case is a solution obvious at first sight.

Appendix B: The Cahn–Hilliard Equation

As mentioned in Sect. 4.3 the Cahn–Hilliard equation is a model for the evolution of a conserved concentration field C during phase separation. Phase separation might for example describe the change of a binary alloy from a uniform mixed state to that of a spatially separated two-phase structure. Such a change occurs, if the temperature of the system is reduced rapidly below a critical temperature T. Comparing the Gibbs free energy $\mathcal{G}$ belonging to a two-phase structure to that of a mixed state results in the phase diagram depicted schematically in Fig. 4.3. The so-called "spinodal" and "conodal"

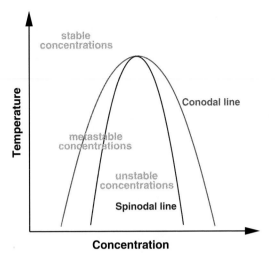

Fig. 4.3. Phase diagram indicating spinodal line (inner curve) and conodal line (outer curve). The lines separate regimes of stable, unstable and metastable concentration for a system undergoing phase separation.

lines separate regimes of stable, unstable and metastable concentrations, respectively.

In Fig. 4.4 the temporal evolution of a binary alloy into a separated two-phase structure starting from a mixed state is displayed. The red and green coloring indicates, which of the two phases is dominant. Taking into account the actual value of the dominant phase, the blurred phase transition and the intensity of the coloring give an impression of the state of the separation.

Appendix C: The Allen–Cahn Equation

In contrast to the Cahn–Hilliard equation, the Allen–Cahn equation describes the evolution of a *non*-conserved order parameter field. This applies for example to the growth of grains in a crystalline sample. If we consider the sample to be composed of several grains, then surface motion does not necessarily have to conserve the amount of material within each individual grain. The total energy of the system would be the sum of the surface energies of the grain interfaces plus the bulk energies of the phases, i.e.:

$$F = \sum_{\alpha\beta} \int_{\mathbf{x} \in S_{\alpha\beta}} \gamma_{\alpha\beta}(n_{\alpha\beta}(\mathbf{x})) \mathrm{d}A + \sum_{\alpha} \int_{V_{\alpha}} \mathcal{F}_{\alpha} \mathrm{d}V \ . \qquad (4.111)$$

At grain junctions usually several equilibrium configurations are possible depending on the shape of the grains. Assuming all grains to be regular octagons, three different equilibrium configurations might arise (Fig. 4.5).

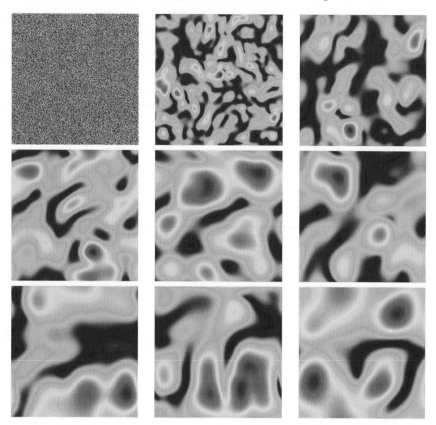

Fig. 4.4. Phase separation of a binary alloy as modeled by the Cahn–Hilliard equation (4.44). The single pictures within the figure represent consecutive states of the system during its temporal evolution from a mixed state (upper left picture) towards a separated two-phase structure (lower right picture). Each snapshot depicts spatial domains for which the two colors red and blue indicate a dominance of either of the two phases. The transition occurs smoothly via yellow and green with respective shading.

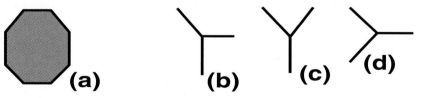

Fig. 4.5. Equilibrium configurations at grain junctions picturing all grains as regular octagons. (a) Geometry of the grain itself. (b)–(d) The three possible configurations at the junction points for this underlying grain geometry. From point of view of energetics other configurations as $\prec$ are identical to the ones depicted in (b)–(d) due to the parity symmetry of the system.

The number of equilibrium triple junctions is even larger if grains are assumed to be regular hexagons. The six possible configurations belonging to this case are depicted in Fig. 4.6.

To determine the temporal evolution of such interface configurations one can employ the following variational formulation: The minimum of

$$\langle \nabla F, \mathbf{v} \rangle + \frac{1}{2\Delta t} \langle \mathbf{v}, \mathbf{v} \rangle \tag{4.112}$$

occurs at $\mathbf{v} = -\nabla F \Delta t$. Moreover, the energy change $F(S_{\mathbf{v}}) - F(S)$ for moving an initial surface configuration S to a new surface configuration $S_{\mathbf{v}}$ is well approximated by $\langle \nabla F, \mathbf{v} \rangle$. Thus one can obtain an iteration for the evolution of the interface configuration by minimizing

$$F(S_{\mathbf{v}}(0)) - F(S(0)) + \frac{1}{2\Delta t} \langle \mathbf{v}, \mathbf{v} \rangle \ . \tag{4.113}$$

Here $S(0)$ is an arbitrary initial configuration. This procedure has to be applied to the configurations $S(n \cdot \Delta t)$ ($n = 1, 2, ...$) until convergence is obtained.

Now given a line segment S_i within a two-dimensional interface configuration $S_{\alpha\beta}$, the energy change for moving segment S_i in normal direction over a distance v_i is given by $v_i(f_{i+} - f_{i-})$. Here $f_{i\pm}$ can be obtained from geometrical considerations. This is illustrated in Fig. 4.7, where $\mathbf{n} = \mathbf{n}(S_i)$ refers to the normal of line segment S_i and l_i to its length. Formally one

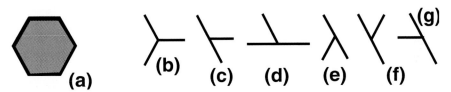

Fig. 4.6. Equilibrium configurations at grain junctions picturing all grains as regular hexagons. (a) Geometry of the grain itself. For this kind of underlying grain geometry six different configurations at junction points are possible. They are displayed in graphs (b)–(g). Again other configurations as $\prec$ can be mapped onto the ones depicted here due to the parity symmetry of the system.

may write

$$f_{i\pm}(S_i) = \delta_{i\pm} \left(\gamma_{\alpha\beta}(\mathbf{n}_{i\pm}) - \mathbf{n}\mathbf{n}_{i\pm} \gamma_{\alpha\beta}(\mathbf{n}) \right) / \sqrt{1 - (\mathbf{n}\mathbf{n}_{i\pm})^2} \ . \tag{4.114}$$

Thus the expression to minimize is

$$\sum_i \left(v_i(f_{i+} - f_{i-}) - \Delta \mathcal{F}_i l_i + \frac{1}{2\Delta t} \frac{v_i^2 \cdot l_i}{M(\mathbf{n}(S_i))} \right) \ . \tag{4.115}$$

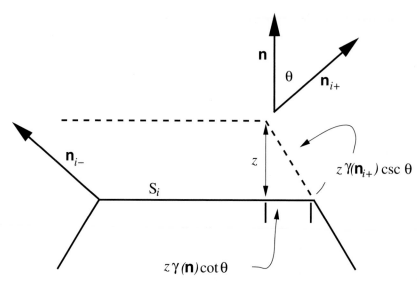

Fig. 4.7. Sketch of how the geometry at a junction point determines the contributions to the energy configuration of a line segment S_i given by $f_{i+} - f_{i-}$. Here the corner labeled by $\mathbf{n}_{i+}$ is depicted. Thus the figure displays the geometric quantities applying to the calculation of f_{i+}. In a similar manner f_{i-} can be obtained from an analogous construction at the left grain corner of line segment S_i. f_{i+} is determined from the line tension in direction of $\mathbf{n}_{i+}$, i.e. $\gamma_{\alpha\beta}(\mathbf{n}_{i+})$, minus the line tension $\gamma_{\alpha\beta}(\mathbf{n})$ projected in direction of $\mathbf{n}_{i+}$, i.e. $\gamma_{\alpha\beta}(\mathbf{n})\mathbf{n}\mathbf{n}_{i+}$, normalized with respect to $\sqrt{1 - (\mathbf{n}\mathbf{n}_{i+})^2}$ (see (4.114)).

Carrying out the minimization results in

$$v_i = M(\mathbf{n}(S_i)) \left(\Delta \mathcal{F}_i - \frac{f_{i+} - f_{i-}}{l_i} \right) \Delta t . \tag{4.116}$$

One recognizes its equivalence to the Allen–Cahn equation (4.45) by identifying v_i in the final expression (4.116) with $\frac{\partial \eta}{\partial t}$ in (4.45) and realizing that $\frac{\partial \eta}{\partial t} = -\nabla F$. This in turn implies that the Allen–Cahn equation contains all of the features of non-conserved grain growth discussed above.

5. Asymptotic Analysis

The previous chapter was devoted to a detailed discussion of the thermo-dynamic concepts underlying the formalism of diffuse interface models. This chapter, on the other hand, introduces a further method required to employ diffuse interface modeling to gain new insight into interfacial growth phe-nomena: This is the method of *matched asymptotic expansion*. Classically this method is employed to establish the equivalence of sharp interface and diffuse interface models and thus is an important concept to validate diffuse interface models. I introduce the method of matched asymptotic expansion in Sect. 5.1 within the general mathematical framework of *matching* to pro-vide a comprehensive background for its application in the context of diffuse interface modeling. This section is succeeded by the discussion of a diffuse interface model for thin film epitaxial growth, which serves as example for the use of the asymptotic matching method purely for validation purposes.

On the other hand, as mentioned in the introductory chapter, an asymp-totic analysis can also serve to find or at least reconsider sharp interface models of growth phenomena for cases in which the latter are not well estab-lished. A necessary requirement to employ diffuse interface models for this task is to ensure thermodynamic consistency of the diffuse interface model. Usually this is an additional step yielding further restrictions after the diffuse interface model has been established via a variational procedure as described in Chap. 4. In this context an analysis ensuring thermodynamic consistency automatically would be very desirable. Only lately such a *generalized* analysis could be formulated for a restricted class of interfacial growth problems [102]. It is described in detail in Sect. 5.3. Finally, in the discussion section at the end of this chapter, the methods presented here are placed into the overall context of diffuse interface modeling.

5.1 A Formal Mathematical Approach Towards Matching

The following is a mathematical excursion about matching. It contains the precise mathematical definition of matching and thereby provides the formal framework of the method of *matched asymptotic expansion*. As indicated in the introductory chapter within the context of diffuse interface modeling only recently one has considered the relevance of higher order matching conditions. From point of view of the underlying mathematical theory obtaining relevant matching conditions does not only depend on determining relevant orders of the expansions, but also on the construction of extended domains of validity. In the following this issue is discussed in its formal mathematical framework and illustrated by hand of a simple example. The discussion of this formal framework follows [31, 115, 197]. Its connection to its application within the context of matched asymptotic expansion as necessary for the analysis of diffuse interface models is given in Sect. 5.1.2.

Here I start from the assumption that $D = [0, 1]$, $I = (0, \varepsilon_1)$ and $y(x, \varepsilon)$ are continuous on $D \times I$. Furthermore, I suppose that there are $y_k(x)$ such that the outer expansion

$$y \sim y_0(x) + \varepsilon y_1(x) + \varepsilon^2 y_2(x) + \cdots \qquad (5.1)$$

is uniformly valid on $[\bar{x}, 1]$, $\bar{x} > 0$, as $\varepsilon \to 0^+$. The inner variable reads

$$X = \frac{x}{\varepsilon} . \qquad (5.2)$$

Moreover, I assume $Y_k(X)$ to exist such that

$$y \sim Y_0(X) + \varepsilon Y_1(X) + \varepsilon^2 Y_2(X) + \cdots \qquad (5.3)$$

uniformly on $[0, \bar{X}]$ for some $\bar{X} < 1/\varepsilon_1$ as $\varepsilon \to 0^+$. This is identical to the assumption that there is both an inner and an outer expansion for the same function $y(x, \varepsilon)$. The function $y(x, \varepsilon)$ can be viewed as a solution of the algebraic problem

$$f(x, y, \varepsilon) = 0 \qquad (5.4)$$

or a boundary-value problem like

$$L_\varepsilon[y] \equiv \varepsilon \frac{d^2y}{dx^2} + a(x)\frac{dy}{dx} + b(x) = f(x, \varepsilon) \qquad \text{with} \qquad x \in (0, 1) , \qquad (5.5)$$
$$y(0, \varepsilon) = A \qquad \text{and} \qquad y(1, \varepsilon) = B . \qquad (5.6)$$

For the "outer limit" in (5.1), x is fixed. For the "inner limit" in (5.3), X is fixed.

Now let $D_o(\bar{x})$ and $D_i(\bar{X})$ denote the regions of uniform validity for the outer and inner expansions, respectively:

$$D_o(\bar{x}) = \{(x, \varepsilon) : x \in [\bar{x}, 1], \varepsilon \in I\} \qquad (5.7)$$
$$D_i(\bar{X}) = \{(x, \varepsilon) : x \in [0, \bar{X}\varepsilon], \varepsilon \in I\} . \qquad (5.8)$$

By definition $D_o(\bar{x})$ and $D_i(\bar{X})$ depend on the ε-independent, fixed values $\bar{x}$ and $\bar{X}$. The precise values proof to be irrelevant for the discussion of overlap regions and matching. Thus I denote these regions simply by D_o and D_i. This underlying picture of D_o and D_i is important to understand the following *extension theorem* necessary to define matching. Extension theorems are theorems which extend the region of uniformity of asymptotic statements like (5.1). An early theorem is due to Kaplun [161]:

Theorem 1 *Let $D = [0, 1]$, $I = (0, \varepsilon_1)$ and $y(x, \varepsilon)$ be continuous on $D \times I$. Also, let $y_0(x)$ be some continuous function on $(0, 1]$ such that*

$$\lim_{\varepsilon \to 0^+} [y(x, \varepsilon) - y_0(x)] = 0 \tag{5.9}$$

uniformly on $[\bar{x}, 1]$ for every $\bar{x} > 0$. Then there exists a function $0 < \delta(\varepsilon) \ll 1$ such that

$$\lim_{\varepsilon \to 0^+} [y(x, \varepsilon) - y_0(x)] = 0 \tag{5.10}$$

uniformly on $[\delta(\varepsilon), 1]$.

(See [99] for further extension theorems.) Obviously there is a whole range of functions satisfying this theorem. One simple example is

$$y(x, \varepsilon) = x + e^{x/\varepsilon} + \varepsilon \quad , \quad y_0(x) = x \ . \tag{5.11}$$

The limit (5.9) implies $y(x, \varepsilon) \sim y_0(x) + \mathcal{O}(1)$ uniformly on $[\bar{x}, 1]$.

Basically the effect of the Kaplun extension theorem is to extend the region of uniform validity D_o to a *layer* around this region, which I call $\hat{D}_o$ in the following. To more carefully define $\hat{D}_o$, *intermediate variables* need to be introduced. Let $\eta(\varepsilon)$ be any function with $0 < \eta(\varepsilon) \ll 1$. Here I define the intermediate variable x_η by

$$x = \eta(\varepsilon) x_\eta \ . \tag{5.12}$$

Then, in accordance to Theorem 1, one may state

$$\lim_{\varepsilon \to 0^+, x_\eta \text{ fixed}} [y(\eta x_\eta, \varepsilon) - y_0(\eta x_\eta)] = 0 \tag{5.13}$$

uniformly on $x_\eta \in [\bar{x}_\eta, 1]$ for all η with $\delta = \mathcal{O}(\eta)$. Generally, when introducing intermediate variables one views η as satisfying $\delta \ll \eta \ll 1$, though to define $\hat{D}_o$ one usually sets η equal to δ or 1:

$$\hat{D}_o(\bar{x}_\eta) = \{(x, \varepsilon) : x \in [\bar{x}_\eta \delta(\varepsilon), 1], \varepsilon \in I\} \ . \tag{5.14}$$

In (5.11), for example, there is an intermediate variable x_η with

$$y(x, \varepsilon) - y_0(x) = e^{\frac{-x_\eta \eta}{\varepsilon}} + \varepsilon = \mathcal{O}(1) \tag{5.15}$$

uniformly on $[\bar{x}_\eta, 1]$ provided $\bar{x}_\eta > 0$ and $\varepsilon \ll \eta$. Here one could for example choose $\delta(\varepsilon) = \varepsilon^{1/2}$ in accordance to the theorem.

In an analogous manner one can construct an extended domain of validity $\hat{D}_i$ for the inner expansion (5.3) choosing as inner variable

$$X = \frac{\eta(\varepsilon)x_\eta}{\varepsilon} .$$
(5.16)

For some (x, ε) near $(0, 0)$ the non-extended domains D_o and D_i do not *overlap*, i.e. they do not intersect regardless of the choices of $\bar{x}$ and $\bar{X}$. In a similar fashion one can construct *non-overlapping extended domains* and *overlapping extended domains*. If there is an overlapping extended domain, there are functions $\eta_i(\varepsilon)$ and $\eta_o(\varepsilon)$ such that for any intermediate variable x_η with $\eta_i(\varepsilon) \ll \eta(\varepsilon) \ll \eta_o(\varepsilon)$ inner and outer expansion are uniformly valid. This implies that given any η with $\eta_i(\varepsilon) \ll \eta(\varepsilon) \ll \eta_o(\varepsilon)$, there is an ε-independent interval I_η such that both

$$\lim_{\varepsilon \to 0^+, x_\eta \text{ fixed}} [y(\eta x_\eta, \varepsilon) - y_0(\eta x_\eta)] = 0$$
(5.17)

$$\lim_{\varepsilon \to 0^+, x_\eta \text{ fixed}} \left[y(\eta x_\eta, \varepsilon) - Y_0 \left(\frac{\eta x_\eta}{\varepsilon} \right) \right] = 0$$
(5.18)

are uniformly on $x_\eta \in I_\eta$, $x_\eta > 0$. Subtracting these expressions one obtains a matching condition:

$$\lim_{\varepsilon \to 0^+, x_\eta \text{ fixed}} \left[y_0(\eta x_\eta) - Y_0 \left(\frac{\eta x_\eta}{\varepsilon} \right) \right] = 0 .$$
(5.19)

Further, if $y_0(0^+)$ and $Y_0(\infty)$ does exist, then

$$\lim_{x \to 0^+} y_0(x) = \lim_{X \to \infty} Y_0(X) .$$
(5.20)

This is the so-called *Prandtl matching condition*. If (5.19) were satisfied, then one would consider the leading-order outer expansion $y_0(x)$ to match the leading-order inner expansion $Y_0(X)$ on an overlap domain

$$\bar{D}_0 = \{(x, \varepsilon) : x_\eta = x\eta \in I_\eta, \eta_i(\varepsilon) \ll \eta(\varepsilon) \ll \eta_o(\varepsilon)\} .$$
(5.21)

At this stage it is worthwhile reflecting a few points in detail. First, $y_0(0^+)$ or $Y_0(\infty)$ may not exist at all. In this case inner and outer expansion cannot be matched to leading-order by use of the Prandtl matching condition. However, it may still be possible to match the expansions by demonstrating the existence of an overlap domain for which (5.19) is satisfied. Second, even if matching condition (5.19) cannot be satisfied, there is still the possibility that an order P of the outer expansion matches an order Q of the inner expansion. That implies that there may be some overlap domain where

$$\lim_{\varepsilon \to 0^+, x_\eta \text{ fixed}} \left[\sum_{n=0}^{P} \varepsilon^n y_n(x_\eta \eta) - \sum_{n=0}^{Q} \varepsilon^n Y_n \left(\frac{x_\eta \eta}{\varepsilon} \right) \right] = 0 .$$
(5.22)

This allows me to define matching:

Definition *Choose $x_\eta = \frac{x}{\eta(\varepsilon)} \in \mathbb{R}$ and let R be any non-negative integer. Consider outer and inner expansion defined in (5.1)- (5.3) to* **match** *to $\mathcal{O}(\varepsilon^R)$ on a common domain of validity $\bar{D}_R(x_\eta)$, if there are functions η_1 and η_2 with $\eta_1 \ll \eta_2$ and integers P, Q such that*

$$\lim_{\varepsilon \to 0^+, x_\eta \text{ fixed}} \frac{M_{PQ}}{\varepsilon^R} =$$

$$\lim_{\varepsilon \to 0^+, x_\eta \text{ fixed}} \left[\frac{\sum_{n=0}^{P} \varepsilon^n y_n(x_\eta \eta) - \sum_{n=0}^{Q} \varepsilon^n Y_n\left(\frac{x_\eta \eta}{\varepsilon}\right)}{\varepsilon^R} \right] = 0 \quad (5.23)$$

for any function η satisfying $\eta_1 \ll \eta \ll \eta_2$ and

$$\bar{D}_R(x_\eta) = \{(x,\varepsilon) : x_\eta = x\eta, \eta_1(\varepsilon) \ll \eta(\varepsilon) \ll \eta_2(\varepsilon)\} . \quad (5.24)$$

5.1.1 Illustration of Extended Domains

To illustrate the formalism developed in the previous part of this section here I consider the following simple example, where I will use the following assumptions throughout the discussion: If $0 < \delta(\varepsilon)$, $x > 0$

$$|\log(\varepsilon)| \ll \delta \quad \Rightarrow e^{-\delta} \ll \varepsilon^n \quad \forall n > 0 \quad (5.25)$$

$$\delta = \mathcal{O}_s(|\log(\varepsilon)|) \Rightarrow e^{-\delta} = \mathcal{O}_s(1) \quad (5.26)$$

$$x \ll \varepsilon |\log(\varepsilon)| \quad \Rightarrow e^{-x/\varepsilon} \ll \varepsilon^n \quad \forall n > 0 . \quad (5.27)$$

Here $\phi = \mathcal{O}_s(\psi)$ denotes $\phi = \mathcal{O}(\psi)$ and $\psi = \mathcal{O}(\phi)$.

In particular I will consider matching of inner and outer expansion of the function

$$y(x,\varepsilon) = \frac{1}{\sqrt{1-4\varepsilon}} \left\{ \exp\left[-(1-\sqrt{1-4\varepsilon})\frac{x}{2\varepsilon}\right] - \exp\left[-(1+\sqrt{1-4\varepsilon})\frac{x}{2\varepsilon}\right] \right\} \quad (5.28)$$

which is the solution of the singular initial value problem

$$\varepsilon y'' + y' + y = 0 , \quad (5.29)$$

$$y(0,\varepsilon) = 0 \quad , \quad y'(0,\varepsilon) = \frac{1}{\varepsilon} . \quad (5.30)$$

Using (5.1) the first two terms of the outer expansion can easily be determined from (5.28). Fixing x and expanding in ε one finds

$$y = (1 + 2\varepsilon + \mathcal{O}(\varepsilon^2)) \left[e^{-x-\varepsilon x + \mathcal{O}(\varepsilon^2)} - e^{-x/\varepsilon + x + \mathcal{O}(\varepsilon)} \right] , \quad (5.31)$$

from which I deduce

$$y_0(x) = e^{-x} \quad , \quad y_1(x) = (2-x)e^{-x} . \tag{5.32}$$

Similarly, to compute the inner expansion by the help of (5.3) one expresses (5.28) in terms of X, fixes X and expands in ε:

$$Y = (1 + 2\varepsilon + \mathcal{O}(\varepsilon^2)) \left[e^{-\varepsilon X - \varepsilon^2 X + \mathcal{O}(\varepsilon^3)} - e^{-X + \varepsilon X + \mathcal{O}(\varepsilon^2)} \right] . \tag{5.33}$$

Thereby one obtains

$$Y_0(X) = 1 - e^{-X} \quad , \quad Y_1(X) = (2-X) - (2+X)e^{-X} . \tag{5.34}$$

Before finding the overlap domains where outer and inner expansion match to $\mathcal{O}(1)$ and $\mathcal{O}(\varepsilon)$, I will discuss how these expansions would arise if one had not known the exact solution *a priori*.

By substituting the outer expansion into (5.29) one obtains the problems:

$$\mathcal{O}(1) : y_0' + y_0 = 0 \tag{5.35}$$
$$\mathcal{O}(\varepsilon) : y_1' + y_1 = -y_0'' . \tag{5.36}$$

Assuming a_0 and b_0 to be constant the general solutions for (5.35) and (5.36) are

$$y_0(x) = a_0 e^{-x} \tag{5.37}$$
$$y_1(x) = (b_0 - a_0 x)e^{-x} . \tag{5.38}$$

Clearly, a_0 cannot be chosen such that $y_0(x)$ satisfies both initial conditions. Therefore, there must be a layer of extended validity at $x = 0$. In terms of Y and X the initial value problem (5.29)–(5.30) can be written as

$$Y'' + Y' + \varepsilon Y = 0 , \tag{5.39}$$

$$Y(0, \varepsilon) = 0 \quad , \quad Y'(0, \varepsilon) = 1 , \tag{5.40}$$

from which one obtains the inner equations

$$\mathcal{O}(1) : Y_0'' + Y_0' = 0 \quad , \quad Y_0(0) = 0 \quad , \quad Y_0'(0) = 1 \tag{5.41}$$
$$\mathcal{O}(\varepsilon) : Y_1'' + Y_1' = -Y_0 \quad , \quad Y_1(0) = 0 \quad , \quad Y_1'(0) = 0 . \tag{5.42}$$

Their solutions are given in (5.34). In contrast to boundary value problems, the unknown constants of integration to be determined from matching are part of the outer solution. If one applied Prandtl matching to match y_0 and Y_0 one would obtain

$$\lim_{x \to 0^+} y_0(x) = a_0 = 1 = \lim_{X \to \infty} Y_0(X) \tag{5.43}$$

and thus recover $y_0(x)$ in (5.32).

To find an extended domain for the outer expansion one assumes $\eta(\varepsilon) \ll 1$ and searches for $\eta_1(\varepsilon)$ such that $\eta_1(\varepsilon) \ll \eta(\varepsilon)$ implies (5.17) for the intermediate variable

$$x_\eta = \frac{x}{\eta} > 0 . \tag{5.44}$$

Given (5.31), the limit is valid as long as $e^{-x_\eta \eta / \varepsilon} \ll 1$. To assure this, one can choose $\eta_1(\varepsilon) = \varepsilon |\log(\varepsilon)|$. Here the notation $\phi \ll= \psi$ implies either $\phi \ll \psi$ or $\phi = O_s(\psi)$. It follows that $\bar{D}_o$ is an extended domain for the outer expansion as long as η satisfies

$$\eta_{1,0} \equiv \varepsilon |\log(\varepsilon)| \ll \eta \ll= 1 . \tag{5.45}$$

Though for $x_\eta \in I_\eta$ each η defines a different region in the (x, ε)-plane, all that really matters for the limit to vanish is that η satisfies (5.45). So it is common practice to consider the extended domain of a single term of the outer expansion of $y_0(x)$ to be given by (5.45).

To find the extended domain for the first-order outer expansion $y_0(x) + \varepsilon y_1(x)$ one assumes $\eta(\varepsilon) \ll 1$ and searches for an $\eta_{1,1}(\varepsilon)$ such that $\eta_{1,1}(\varepsilon) \ll \eta(\varepsilon)$ implies

$$\lim_{\varepsilon \to 0^+, x_\eta \text{ fixed}} \frac{[y(\eta x_\eta, \varepsilon) - y_0(\eta x_\eta) - \varepsilon y_1(\eta x_\eta)]}{\varepsilon} = 0 . \tag{5.46}$$

Again from (5.31) one obtains that the above limit holds if η satisfies (5.45), i.e. the choice $\eta_{1,1} = \eta_{1,0}$ works. If one continues this process of extending the domain in an Rth order outer expansion to find $\eta_{1,R}$ it is often the case that $\eta_{1,R} \ll \eta_{1,R+1}$. The reason is that adding further terms to the limit places additional restrictions on η. For this particular example the extended outer domains at $\mathcal{O}(1)$ and $\mathcal{O}(\varepsilon)$ turn out to be the same.

To find an extended domain for the inner expansion to lowest order $\mathcal{O}(1)$ one assumes $\varepsilon \ll \eta(\varepsilon)$ and seeks an $\eta_2(\varepsilon)$ such that $\eta \ll \eta_2(\varepsilon)$ implies (5.18). Again from (5.31) it is easy to verify that the extended domain for this inner expansion is defined by

$$\varepsilon \ll= \eta \ll \eta_{2,0} \equiv 1 . \tag{5.47}$$

Finding the extended inner domain to $\mathcal{O}(\varepsilon)$ is more delicate. In terms of the intermediate variables one obtains

$$\frac{y(\eta x_\eta, \varepsilon)}{\varepsilon} = \frac{1}{\varepsilon} - \frac{e^{-x_\eta \eta / \varepsilon}}{\varepsilon} - \frac{\eta}{\varepsilon} x_\eta - \frac{\eta}{\varepsilon} x_\eta e^{-x_\eta \eta / \varepsilon}$$
$$+ 2 - 2 e^{-x_\eta \eta / \varepsilon} + \mathcal{O}(\eta) + O\left(\frac{\eta^2}{\varepsilon}\right) + \mathcal{O}(\varepsilon) . \tag{5.48}$$

Moreover the intermediate variables yield the expression

$$\frac{1}{\varepsilon}Y_0 + Y_1 = \frac{y(\eta x_\eta, \varepsilon)}{\varepsilon} = \frac{1}{\varepsilon} - \frac{e^{-x_\eta \eta/\varepsilon}}{\varepsilon} - \frac{\eta}{\varepsilon}x_\eta - \frac{\eta}{\varepsilon}x_\eta e^{-x_\eta \eta/\varepsilon} + 2 - 2e^{-x_\eta \eta/\varepsilon} .$$

$$(5.49)$$

Subtracting these two expressions it becomes apparent that

$$\lim_{\varepsilon \to 0^+, x_\eta \text{ fixed}} \frac{[y(\eta x_\eta, \varepsilon) - Y_0(\eta x_\eta/\varepsilon) - \varepsilon Y_1(\eta x_\eta/\varepsilon)]}{\varepsilon} = 0 \qquad (5.50)$$

provided $\eta^2/\varepsilon \ll 1$. Thus the choice $\eta_{2,1} = \varepsilon^{1/2}$ ensures that the limit vanishes and that the extended inner domain to $\mathcal{O}(\varepsilon)$ reads

$$\varepsilon \ll = \eta \ll \eta_{2,1} \equiv \varepsilon^{1/2} . \qquad (5.51)$$

Here the extended domain to $\mathcal{O}(\varepsilon)$ is "smaller" than the domain to $\mathcal{O}(1)$, i.e. $\eta_{2,1} \ll \eta_{2,0}$.

Based on the previous discussion it is obvious that the overlap domains to $\mathcal{O}(1)$ and $\mathcal{O}(\varepsilon)$ are, respectively:

$$\eta_{1,0} \ll \eta \ll \eta_{2,0} \qquad (5.52)$$
$$\eta_{1,1} \ll \eta \ll \eta_{2,1} \qquad (5.53)$$

or

$$\varepsilon |\log(\varepsilon)| \ll \eta \ll 1 \qquad (5.54)$$
$$\varepsilon |\log(\varepsilon)| \ll \eta \ll \varepsilon^{1/2} . \qquad (5.55)$$

Now three cases can be distinguished:

1. If η satisfies these asymptotic relations, outer and inner expansion match to $\mathcal{O}(1)$ and $\mathcal{O}(\varepsilon)$, respectively. In particular if η satisfies (5.52) then

$$\lim_{\varepsilon \to 0^+, x_\eta \text{ fixed}} [y_0(\eta x_\eta) - Y_0(\eta x_\eta/\varepsilon)] = 0 . \qquad (5.56)$$

2. If η satisfies the more stringent requirement (5.53), then

$$\lim_{\varepsilon \to 0^+, x_\eta \text{ fixed}} \frac{[y_0(\eta x_\eta) + \varepsilon y_1(\eta x_\eta) - Y_0(\eta x_\eta/\varepsilon) - \varepsilon Y_1(\eta x_\eta/\varepsilon)]}{\varepsilon} = 0 .$$

$$(5.57)$$

3. If the exact solution y was not known a priori then one would choose a_0 in the incomplete outer solution $y_0(x) = a_0 e^{-x}$ and find $\eta_{1,0}, \eta_{2,0}$ such that (5.56) is satisfied.

Remarks on Overlap Domains and Matching

1. General theorems showing the existence of overlap domains have not been found [196]. In practice, the existence of overlap domains where inner and outer solutions can be matched is done on a case by case basis.

2. Prandtl matching corresponds to lowest-order matching with $P = Q = R = 0$.
3. In some problems P and Q may not be known a priori. Moreover, P may not equal Q.
4. Some expansions cannot be matched. The matching in (5.23) is defined with respect to underlying gauge functions $\phi_n(\varepsilon) = \varepsilon^n, n \geq 0$. Obviously, some functions y may have more general outer expansions:

$$y(x, \varepsilon) \sim \sum_{n \geq 0} \phi_n(\varepsilon) y_n(x) . \tag{5.58}$$

Indeed, the inner variable could be defined in a more general way, $X = x/\delta(\varepsilon), 0 < \delta \ll 1$. Then the inner expansion may be expressed with respect to different gauge functions. However, usually these sorts of generalizations are not taken into account.

5.1.2 Matching in the Context of Diffuse Interface Modeling

In the context of diffuse interface modeling one employs an *asymptotic analysis* to establish the relationship between diffuse interface and sharp interface modeling equations. The basic idea underlying this *asymptotic analysis* is the *method of matched asymptotic expansion*, whereby matching conditions are employed to find integration constants occurring in the inner expansion. In this context an outer solution, valid away from the interface, is matched to an inner solution, which is valid in the interfacial region [115]. The inner expansion near a point $\mathbf{x}_0$ on the interface is described via a stretched coordinate ζ, with $\mathbf{x} = \mathbf{x}_0 + \varepsilon \zeta \hat{\mathbf{n}}$, where $\hat{\mathbf{n}}$ is the local unit vector normal to the interface. The low-order terms in an inner expansion of the solution can be obtained by using the relations

$$\nabla^2 = \frac{1}{\varepsilon^2} \left(\frac{\partial}{\partial \zeta} \right)^2 + \frac{\tilde{\kappa}}{\varepsilon} \frac{\partial}{\partial \zeta} + \mathcal{O}(1) \tag{5.59}$$

and

$$\frac{\partial}{\partial t} = -\frac{\tilde{v}_n}{\varepsilon} \frac{\partial}{\partial \zeta} + \mathcal{O}(1) . \tag{5.60}$$

Here the dimensionless mean curvature $\tilde{\kappa}$ and the dimensionless interface speed $\tilde{v}_n$ are assumed to be of order unity. In the inner region the phase-field and the transport field are expanded in powers of ε,

$$\Phi = \Phi^0 + \varepsilon \Phi^1 + \mathcal{O}(\varepsilon^2) , \tag{5.61}$$

$$c = c^0 + \varepsilon c^1 + \mathcal{O}(\varepsilon^2) . \tag{5.62}$$

The resulting equations are solved order by order in ε, with far-field boundary conditions, which are obtained by matching to the outer solution. The outer solution of the phase-field has $\Phi(\mathbf{x}) = 0$ or $\Phi(\mathbf{x}) = 1$ to all orders on either side of the boundary Γ_{ij}. The transport field has the expansion $C(\mathbf{x}) = C(\mathbf{x})^{(0)} + \varepsilon C^{(1)}(\mathbf{x}) + \mathcal{O}(\varepsilon^2)$. The limiting behavior near $\mathbf{x}_0$ is given by

$$C(\mathbf{x}_0 + \varepsilon \zeta \hat{\mathbf{n}}) = C_{\pm}^{(0)}(\mathbf{x}_0) + \varepsilon[C_{\pm}^{(1)}(\mathbf{x}_0) + \zeta \frac{\partial C_{\pm}^{(0)}}{\partial n}(\mathbf{x}_0)] + \mathcal{O}(\varepsilon^2) , \qquad (5.63)$$

where $C_{+}^{(0)}$ and $C_{-}^{(0)}$ denote the limits of C as $\zeta \to 0_+$ and $\zeta \to 0_-$, respectively, and so forth. The inner solution must match with this behavior as $\zeta \to \pm\infty$ [49, 50, 115]. In this way matching conditions yield the physics of the interfacial region. Typically to dominant order the motion of the diffuse interface is governed by a Stefan problem. The next order approximation brings in effects relating the temperature at the interface to its curvature and normal velocity. Physically such effects arise from surface tension, which is therefore an essential ingredient of diffuse interface models. If corresponding sharp interface models are well established those have to be recovered via the asymptotic analysis. This might impose restrictions to the numerical parameters of the diffuse interface model as will be discussed in the numerical appendix of this book. On the other hand this analysis might also lead to the reconsideration of sharp interface modeling equations (or even completely new formulations of such models) as I will demonstrate in Chap. 6.

5.2 Asymptotic Matching for Thin Film Epitaxial Growth

This section discusses an example [110], which displays how the above asymptotic analysis is employed to validate a diffuse interface model for a free boundary problem. The example chosen is that of *thin film epitaxial growth*. For this phenomenon a set of sharp interface equations constituting a well established free boundary type model does exist. An essential ingredient of this sharp interface model are terms to describe the attachment kinetics at the interface. How to recover the precise kinetics via a diffuse interface model is not obvious at first sight. This section discusses an approach, which derives the diffuse interface equations from two thermodynamic potentials, namely the inner energy functional $\mathcal{E}$ and the entropy functional $\mathcal{S}$. This way, when writing down the free energy density function, an additional degree of freedom compared to models derived from one potential only is introduced. This term serves to recover the precise attachment kinetics in a first order analysis. Thus this section provides essentially an example, in which matching is used purely for the purpose of validation. The first order analysis becomes necessary to recover the full physics of the sharp interface model, in particular the kinetics involved. Since first order analysis is based on a sharp interface analysis, it is an example, which develops the matching analysis in the context of diffuse interface modeling from its classical limit to the so-called thin interface limit. The latter has been introduced in the context of dendritic growth by Karma and Rappel in [168].

5.2.1 Motivation

Growth of thin crystalline films involves the physical transport of material to a surface, where film constituents are incorporated. Appropriate conditions for such growth processes can be established by various kinds of epitaxy. The ultimate goal of epitaxial growth is to control the structuring of a thin film on smaller and smaller scales. Modeling and simulating growth dynamics and kinetics during epitaxy is a useful tool to predict the structures to be expected under certain growth conditions. It contributes to an understanding of what physical mechanisms act on which scales and of what kind of growth morphologies are triggered by their interaction. One of the challenges to physicists and material scientists concerned with this task is the range of length and time scales represented by problems of practical interest, e.g. the growth of device layers [141]. Atomistic processes on the nanoscale as the attachment of single atoms to a step or a defect can significantly affect the evolution of an overall morphology on a surface–subsection several orders of magnitude larger [21]. Moreover, depending on the epitaxial conditions, the precise transport as well as the precise kinetic mechanisms might vary. Growth modes might be as diverse as step flow growth, island growth or layer-by-layer growth [47] with possible transitions from one to the other. A model for epitaxial growth with potential for engineering applications would therefore (a) describe spatial and temporal scales of several orders in magnitude, (b) provide for a straightforward extension to further transport fields and kinetic quantities and (c) be capable of dealing with the topological changes arising from the transition between growth modes.

Focusing just on the surface height profile of a film above a surface, continuum equations in the form of partial differential equations [235, 287, 288] describe the evolution of surface morphologies at large scales. However, a necessary assumption for their applicability is microscopic smoothness of the surface, for the spatial derivatives to exist. Therefore they are unsuitable for the description of roughening on the atomic scale, which is a big concern in many device applications. On the other hand, particle based numerical approaches as Kinetic Monte Carlo (KMC) algorithms can be employed to model the atomistic processes in more detail and thereby constitute an alternative to continuum equations [73, 211, 216, 293]. A big advantage of these models is their capability to quantify atomistic kinetic processes via attachment rates, which can be determined from first principle calculations [242]. They have been applied widely to various issues of epitaxial growth in the past [276] and constitute a comprehensive framework to describe homoepitaxial systems as well as very limited cases of heteroepitaxial growth [105]. However, they are not easily extended to include further growth mechanism as, e.g., hydrodynamic transport in the bulk, elastic interaction between steps or entropic confinement. Moreover, modeling systems of practical interest is not feasible, since simulations are usually based on length and time scales of single atoms and adatom hopping rates.

To meet challenges (a) to (c) formulated above and simulate surface–sectors of sizes of practical interest, continuum models allowing for fast computation on the one hand and for the incorporation of quantitative kinetic information on the atomic scale on the other hand, constitute a promising modeling approach. Their common feature is the formulation of growth problems as the one arising in epitaxy as a *free-boundary problem* either in its sharp interface or its diffuse interface formulation.

5.2.2 Sharp Interface Formulation

Mathematically the problem of epitaxial growth can be understood as the solution of one (or several coupled[1]) transport equations for a variable ϕ in two domains Ω_1 and Ω_2, which are separated by a boundary Γ_{ij} defined by (2.18):

$$\frac{\partial \phi}{\partial t} + \mathbf{u} \cdot \nabla \phi = \nabla \cdot k \nabla \phi + f . \tag{5.64}$$

Equation (5.64) is the most general form of a diffusion–advection equation, in which k is a function of the vector of space coordinates $\mathbf{x}$. $\mathbf{u}$ is the convective speed, if hydrodynamic transport is considered in Ω_1 and/or Ω_2. $\mathbf{u}$ may be a function of ϕ. ϕ itself may be discontinous at the interface, i.e.:

$$[\phi]_{\Gamma_{ij}} = \beta \tag{5.65}$$

and/or

$$[k\frac{\partial \phi}{\partial n}]_{\Gamma_{ij}} = \gamma . \tag{5.66}$$

The notation $[g]_{\Gamma_{ij}} = g^+ - g^-$ denotes the difference of the values taken by the function $g(\mathbf{x})$ in the limits $\mathbf{x} \to 0$ and $-\mathbf{x} \to 0$, if the interface is assumed to be at 0. f is a body–force term. The interface is subject to boundary conditions of the general form:

$$F(\phi, \frac{\partial \phi}{\partial n}) = \alpha \Gamma_{ij} . \tag{5.67}$$

These boundary conditions ensure the conservation of mass and the conservation of energy at the interface during growth.

Equations (5.64)–(5.67) can be adapted to constitute the *two-sided* sharp interface model for epitaxial growth, i.e.:

---

[1] For the sake of simplicity I will stick to the case of one transport field in the following. Extension to two or more is straightforward.

$$\frac{\partial \rho}{\partial t} = D\nabla^2\rho - \frac{\rho}{\tau^{\mathrm{ev}}} + F \tag{5.68}$$

$$D\frac{\partial \rho^+}{\partial n}|_{\Gamma_{ij}} = k^+(\rho^+ - \rho^{\mathrm{eq}})_{\Gamma_{ij}} \tag{5.69}$$

$$-D\frac{\partial \rho^-}{\partial n}|_{\Gamma_{ij}} = k^-(\rho^- - \rho^{\mathrm{eq}})_{\Gamma_{ij}} \tag{5.70}$$

$$v_n = \Omega D(\frac{\partial \rho^+}{\partial n} - \frac{\partial \rho^-}{\partial n})_{\Gamma_{ij}} . \tag{5.71}$$

Here v_n is the normal velocity at the interface. The transport variable ϕ of (5.64) is the density of adatoms above the surface denoted by ρ. The dynamics of the adatoms on the surface is a diffusion–relaxation process as described by (5.68) and visualized in Fig. 5.1 with an external supply of atoms to the surface quantified by F entering the equation, as well. Adatoms may evaporate after a time τ^{ev}. Equations (5.69)–(5.71) are boundary conditions of the general form of (5.65) and (5.66), which ensure the conservation of energy, (5.69) and (5.70), and the conservation of mass, (5.71), at the boundary Γ_{ij}. D is the diffusion constant involved in the process, Ω the atomic area of the atoms under consideration and k^+ and k^- are coefficients to quantify the kinetic attachment to the boundary. They are different depending on whether a boundary is reached from an upper (denoted by $^-$) or lower side (denoted by $^+$). This is due to the Ehrlich–Schwoebel effect [104], which describes an energy barrier the adatom attaching from the upper side has to overcome. Notice that this difference results in two equations for the conservation of energy to be evaluated in "$+$-direction" respective "$-$-direction", namely (5.69) and (5.70). The whole set of equations dates back to Burton, Cabrera and Franck in 1956 [47]. Obviously a boundary can have very different kinds of geometries depending on the epitaxial growth mode of the substrate. Pictured in Fig. 5.1 are boundaries during step flow and island growth. Note that in the case of epitaxial growth anisotropy due to the underlying crystal structure of the material is contained in $\rho^{\mathrm{eq}} = \rho_0^{\mathrm{eq}} \cdot (1 - \beta\cos4\theta)$. Detailed studies of the influence of crystalline anisotropy in epitaxial growth are given in [106, 109].

The set of equations (5.68)–(5.71) constitutes a modeling approach which can be understood as a direct translation of the basic physical mechanisms of epitaxy into mathematical equations. The physical picture consists of transport in the bulk, see (5.68), and attachment kinetics at the interface, see (5.69)–(5.71). Going back to the modeling challenges (a) to (c) formulated in Sect. 5.2.1 it is important to notice that challenge (a), namely the description of scales of several orders in magnitude, can be achieved due to the fact that these continuum equations themselves can be employed to simulate structures of the order of 10^4 Å, while at the same time information on the atomic scale enters the equations via the attachment coefficients k^- and k^+.

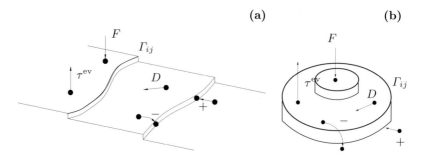

Fig. 5.1. Visualization of kinetics and dynamics during epitaxial growth: Adatoms reach the surface with a rate that is proportional to F. They diffuse on the surface and may desorb from it according to the desorption parameter τ^{ev}. Reaching a boundary Γ_{ij} (e.g. a step (a) or an island (b)) they are incorporated with kinetic attachment rates depending on their position with respect to the surface, i.e. from up or down. This results in growth, i.e. movement of the boundary.

These can be determined on the basis of first-principle calculations.[2] Also (b), namely the extension of the model to further transport fields and kinetic mechanisms, can directly be addressed in this formulation: a second (diffusive) transport field would yield a further diffusion equation to be coupled to (5.68). In the same way hydrodynamic transport could be included in the model via a coupling to the Navier–Stokes equations. Inserting other physical mechanisms as diffusion along steps or elastic step interaction into the model will change the boundary conditions in two ways: On the one hand it results in a change of the chemical potential at the steps and thereby in a different expression for ρ^{eq} in (5.69) and (5.70). On the other hand it enters (5.71) in gradient form to take into account the changed mass balance.

5.2.3 The Diffuse Interface Model Equations and Their Asymptotic Analysis

The basic idea underlying the phase-field type of modeling epitaxial growth is to replace boundary conditions (5.69)–(5.71) of the sharp interface formulation by an additional differential equation for the order parameter Φ introduced in Sect. 4.2. Here I assume Φ to vary smoothly over the boundaries Γ_{ij} such that different layers of the substrate are labeled by different integer numbers (see Fig. 5.2). [3] This convention has been employed to model epitaxial growth by means of a phase-field before by Karma and Plapp [170] as well as by Liu and Metiu [207]. However, in their contributions equal

---

[2] Obviously there is a lower limit to the structures to be resolved which stems from the assumption of a continuum description.

[3] The variation has to take place on a scale much less than the structuring of the substrate.

attachment kinetics at the interface had to be assumed. That way, the problem was directly mapped to that of dendritic growth with equal diffusion constants in both phases. No discontinuities causing fundamental modeling-related questions arose. Here, on the other hand, I propose a diffuse interface model which handles the case of unequal attachment kinetics from both sides of the interface. This model is thermodynamically consistent. It is derived from conservation laws for the energy as well as the entropy. Thereby an additional degree of freedom is obtained. This is one important new feature of the model necessary to recover the full two-sided growth problem described by equations (5.68)–(5.71) in the thin interface limit. The second essential new ingredient for the thin interface analysis to work out correctly is the introduction of a "continuous attachment function" $\tilde{\xi}$, which varies smoothly in the interfacial region, see (5.82) of the following detailed derivation of the model.

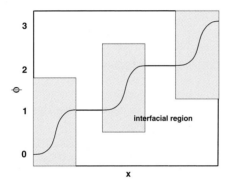

Fig. 5.2. The phase-field function Φ as it varies over several layers of an epitaxial substrate.

As discussed above the following derivation of model equations is based on conservation laws for energy as well as entropy. These may be written in the form

$$\frac{d\mathcal{E}}{dt} + \int_{\Gamma_{ij}} \mathbf{J}_E \cdot \hat{\mathbf{n}} dA = 0 \tag{5.72}$$

$$\frac{d\mathcal{S}}{dt} + \int_{\Gamma_{ij}} \mathbf{J}_S \cdot \hat{\mathbf{n}} dA = \int_V \dot{s}_P dV , \tag{5.73}$$

where $\mathbf{J}_E$ and $\mathbf{J}_S$ represent the flux of energy and entropy through the boundary Γ_{ij}, which has an outward unit normal $\hat{\mathbf{n}}$. $\dot{s}_P$ represents the entropy production term. Following [110] for the inner energy functional $\mathcal{E}$ and the inner entropy functional $\mathcal{S}$ an ansatz of the form

$$\mathcal{E} = \int_V (e(\phi, \rho) + \frac{\xi}{2}|\nabla\phi|^2)dV \tag{5.74}$$

$$\mathcal{S} = \int_V (s(\phi, \rho) + \frac{\xi}{2}|\nabla\phi|^2)dV \tag{5.75}$$

is chosen. The term $\frac{\xi}{2}|\nabla\phi|^2$ denotes the contribution of the diffuse interface, where ξ is a gradient coefficient, which can be assumed constant. e and s are the internal energy and entropy densities, respectively. Their relation to the free energy density is given by

$$s(\phi, \rho) = -\frac{\partial f}{\partial \rho} \quad \text{and}$$

$$e(\phi, \rho) = f(\phi, \rho) + \rho s(\phi, \rho) \ .$$

Thermodynamic consistency requires positive entropy production. Governing equations, which guarantee positive local energy production are obtained by choosing

$$\mathbf{J_E} = -k\nabla\rho - \xi\frac{\partial \phi}{\partial t}\nabla\phi \tag{5.76}$$

$$\mathbf{J_S} = -\frac{k}{\rho}\nabla\rho - \xi\frac{\partial \phi}{\partial t}\nabla\phi \ . \tag{5.77}$$

with a non-classical flux $\xi\frac{\partial \phi}{\partial t}\nabla\phi$ associated with the interfacial region [289]. Entropy production is then given by

$$\rho\dot{s}_P = \frac{\xi}{\rho}|\nabla\rho|^2 + \frac{\partial \phi}{\partial t}[(\xi + \xi\rho)\nabla^2\phi - \frac{\partial f}{\partial \phi}] \ . \tag{5.78}$$

The corresponding phase-field equation is chosen such that the latter expression takes the form $\frac{\xi}{\rho}|\nabla\rho|^2 + (\frac{\partial \phi}{\partial t})^2 \cdot \tau$ with $\tau > 0$. Thus considering a free energy of the form

$$f(\phi, \rho) = \frac{V_E + \rho V_S}{2}d(\phi) + \frac{(\rho - \rho^{eq})}{\rho^{eq}}k(\phi) - \rho\ln(\frac{\rho}{\rho^{eq}}) + \rho \tag{5.79}$$

the following set of governing equations can be derived for the case of epitaxial growth:

$$\tau\frac{\partial \phi}{\partial t} = (\xi + \rho\xi)\nabla^2\phi - \frac{\partial f}{\partial \phi}$$

$$= (\xi + \rho\xi)\nabla^2\phi - \frac{V_E + \rho V_S}{2}d'(\phi) - \frac{\rho - \rho^{eq}}{\rho^{eq}}k'(\phi) \ , \tag{5.80}$$

$$\frac{\partial \rho}{\partial t} = D\Omega\nabla^2\rho + \mathcal{G}(\phi)\frac{\partial \phi}{\partial t} + F + \frac{\rho}{\tau^{ev}} \ . \tag{5.81}$$

Here $\tau^{-1} > 0$ denotes a mobility coefficient, $d(\phi) = \phi^2(1 - \phi)^2$ a double–well potential function with a barrier height given by $V_E + \rho V_S$. $(\xi + \rho\xi) > 0$

serves as gradient energy coefficient and $k(\phi) = \phi^2(3 - 2\phi)$. The source term $\mathcal{G}(\phi) \cdot \frac{\partial \phi}{\partial t}$ describes the influence of movement of the boundary Γ_{ij} on the evolution of the transport field ρ, with $\mathcal{G}(\phi) = \xi \nabla^2 \phi - d'(\phi)\frac{\xi}{2} + 1$. Within $\mathcal{G}(\phi)$ the function $\tilde{\xi}$ in its suitably rescaled form is given by:

$$\tilde{\xi} = -\frac{D}{k^-}\frac{12\mathcal{X}}{l}(1 - \phi) + \frac{D}{k^+}\frac{12\mathcal{X}}{l}(\phi) . \tag{5.82}$$

The rescaling of the model equations simplifies the *asymptotic analysis* by which the sharp interface equations (5.68)–(5.71) are to be recovered. It is this asymptotic analysis, which serves as validation for the model at this point. I will turn to it next.

Equation (5.80) and (5.81) can be made dimensionless by employing the macroscopic length scale $\mathcal{X}$, the time scale $\frac{\mathcal{X}}{D}$ and a scale for the governing field ρ set to unity. Moreover, l defines the interface width, i.e. the scale on which the phase field ϕ varies from one integer value to another. l_c takes the role of a capillary length, which is proportional to surface tension γ, i.e.:

$$l_c = \frac{\gamma \Omega}{k_B T} , \tag{5.83}$$

where k_B refers to the Boltzmann constant and T to the temperature. With abbreviations $\xi^{\text{eq}} = \xi + \rho^{\text{eq}}\xi$ and $V^{\text{eq}} = V_E + \rho^{\text{eq}}V_S$ these terms allow me to define the following dimensionless parameters:

$$\varepsilon = \frac{l}{\mathcal{X}} \quad , \quad \Psi = \frac{\rho^{\text{eq}}D \cdot \tau}{6l \cdot l_c} \quad , \quad \lambda = \frac{l}{6 \cdot l_c} \quad ,$$

$$\nu = \frac{V_s}{V^{\text{eq}}} \quad , \quad \gamma = \frac{\xi}{\xi^{\text{eq}}} . \tag{5.84}$$

Based on those dimensionless parameters equations (5.80) and (5.81) can be formulated as:

$$\varepsilon^2 \Psi \frac{\partial \phi}{\partial t} = \varepsilon^2(1 + \gamma c)\nabla^2 \phi - \frac{(1 + \nu c)}{2}d'(\phi) - \lambda c k'(\phi) , \tag{5.85}$$

$$\frac{\partial c}{\partial t} = \tilde{D}\tilde{\Omega}\nabla^2 c + \tilde{\mathcal{G}}(\phi)\frac{\partial \phi}{\partial t} - c , \tag{5.86}$$

with the rescaled $\tilde{\mathcal{G}}(\phi) = \varepsilon^2 \tilde{\xi}\nabla^2 \phi - \frac{\mu}{2}d'(\phi) + 1$.

To validate the above model equations one employs the *method of matched asymptotic expansion* (see Sect. 5.1.2) with relations (5.59) and (5.60) for the low-order terms in the inner expansion. To keep the succeeding analysis clear for $(\frac{\partial}{\partial \zeta})^2$ the form $\partial_{\zeta\zeta}$ is used, for $\frac{\partial}{\partial \zeta}$ the form ∂_ζ and ∂_t denotes $\frac{\partial}{\partial t}$. Applying the method of matched asymptotic expansion to the dimensionless model equations (5.85)–(5.86) yields a leading-order phase-field equation:

$$\phi_{\zeta\zeta}^{(0)} - \frac{1}{2}d'(\phi^{(0)}) = 0 . \tag{5.87}$$

The leading-order equation for the transport field reads:

$$\tilde{D}\tilde{\Omega}c_{\zeta\zeta}^{(0)} = 0 \,. \tag{5.88}$$

Integrating and matching the outer solution twice in succession gives that $c^{(0)} = C_+^{(0)}(\mathbf{x}_0) = C_-^{(0)}(\mathbf{x}_0)$ is constant. The first-order phase-field equation follows as:

$$\phi_{\zeta\zeta}^{(1)} - \frac{1}{2}g''(\phi^{(0)}\phi^{(1)}) = -(\psi\tilde{v}_n + \tilde{\kappa})\phi_\zeta^{(0)} + \Lambda c^{(0)}k'(\phi^{(0)}) \,, \tag{5.89}$$

and a solvability condition for this equation is given by

$$c^{(0)} = d_0\kappa + \tilde{\beta}\frac{v_n}{v_0} \,, \tag{5.90}$$

where v_0 corresponds to the velocity of a straight step. This can be identified as a modified Gibbs–Thomson boundary condition. With $\tilde{\beta} \to 0$ it can be understood as an equilibrium concentration corresponding to ρ^{eq} in equations (5.69) and (5.70). Thus in this 0th order of the expansion the concentrations right at the interface, denoted by ρ^+ and ρ^- in (5.69) and (5.70), would simply be ρ^{eq}. The difference to this equilibrium value is of kinetic origin and retrieved in the next order of the expansion.

For the field variable a first-order equation reads:

$$\tilde{v}_n\phi_\zeta^{(0)} = \tilde{D}\tilde{\Omega}c_{\zeta\zeta}^{(1)} \,. \tag{5.91}$$

By integrating this equation over the interval $-\infty < \zeta < \infty$ and using the matching conditions, the leading-order mass–conservation condition is recovered in the dimensionless form

$$\tilde{v}_n = \tilde{D}\tilde{\Omega}\frac{\partial C_+^{(0)}}{\partial n} - \tilde{D}\tilde{\Omega}\frac{\partial C_-^{(0)}}{\partial n} \,. \tag{5.92}$$

Thus the boundary condition (5.71) of the sharp interface formulation is retrieved. It still remains to be proven that the boundary equations (5.69) and (5.70) can be obtained in this order of the expansion, as well.

Following the procedure in [110], the leading- and first-order phase-field equations are obtained as

$$\phi_{\zeta\zeta}^{(0)} - \frac{1}{2}d'(\phi^{(0)}) = 0 \,, \tag{5.93}$$

$$\phi_{\zeta\zeta}^{(1)} - \frac{1}{2}d''(\phi^{(0)})\phi^{(1)} = \tilde{r}_1 \,, \tag{5.94}$$

where

$$\tilde{r}_1 = -\gamma c^{(1)}\phi_{\zeta\zeta}^{(0)} + \frac{\nu}{2}c^{(1)}d'(\phi^{(0)}) - (\tau\tilde{v}_n + \tilde{\kappa})\phi_\zeta^{(0)} + \lambda c^{(1)}k'(\phi^{(0)}) \,. \tag{5.95}$$

The first two terms of $\tilde{r}_1$ can be simplified using (5.93) to yield:

$$-\gamma c^{(1)}\phi_{\zeta\zeta}^{(0)} + \frac{\nu}{2}c^{(1)}d'(\phi^{(0)}) = \frac{\tilde{\nu}}{2}c^{(1)}g'(\phi^{(0)})\,, \tag{5.96}$$

where $\tilde{\nu}$ is defined as $\nu + \gamma$. The solvability condition for (5.94) can then be written in the form

$$0 = \int_{-\infty}^{\infty} \phi_{\zeta}^{(0)}\tilde{r}_1 d\zeta = \lambda \int_{-\infty}^{\infty} c^{(1)}k'(\phi^{(0)})\phi_{\zeta}^{(0)} d\zeta$$

$$+ \frac{\tilde{\nu}}{2}\int_{-\infty}^{\infty} c^{(1)}d'(\phi^{(0)})\phi_{\zeta}^{(0)} d\zeta - \frac{1}{6}(\psi\tilde{v}_n + \tilde{\kappa})\,, \tag{5.97}$$

which involves the first-order transport field solution $c^{(1)}$.

With $c^{(1)} = 0$ the transport field equation is satisfied identically to leading-order. The first-order transport field equation reads

$$\tilde{v}_n\phi^{(0)}\phi_{\zeta}^{(0)} - \frac{\mu}{2}\tilde{v}_nd'(\phi^{(0)})\phi_{\zeta}^{(0)} = \tilde{D}\tilde{\Omega}c_{\zeta\zeta}^{(1)} - \tilde{\xi}\tilde{v}_n\phi_{\zeta}^{(0)}\phi_{\zeta\zeta}^{(0)}\,, \tag{5.98}$$

which, when simplified by using (5.93), can be integrated to result in

$$\tilde{D}\tilde{\Omega}c_{\zeta}^{(1)} - \tilde{v}_n\phi(0) - \frac{\tilde{\xi}_1\tilde{v}_n}{2}(\phi_{\zeta}^{(0)})^2 = \tilde{D}\tilde{\Omega}\frac{\partial C_-^{(0)}}{\partial n} = \tilde{D}\tilde{\Omega}\frac{\partial C_+^{(0)}}{\partial n} - \tilde{v}_n\,. \tag{5.99}$$

Here $\tilde{\xi}_1 = \tilde{\xi} - 1$. The constant of integration has been evaluated in two ways by taking $\zeta \to \pm\infty$ and applying the matching conditions (5.63). The solution can then be expressed in the equivalent forms

$$c^{(1)}(\zeta) = C\,{}^{+(1)}(\mathbf{x}) + \zeta\frac{\partial C_+^{(0)}}{\partial n} - \int_{\zeta}^{\infty} \frac{1}{\tilde{\Omega}\tilde{D}}\left(\tilde{v}_n\phi^{(0)} + \frac{\tilde{\xi}_1\tilde{v}_n}{2}(\phi_{\zeta}^{(0)})^2\right)d\eta\,, \tag{5.100}$$

$$c^{(1)}(\zeta) = C_-^{(1)}(\mathbf{x}) + \zeta\frac{\partial C_-^{(0)}}{\partial n} - \int_{-\infty}^{\zeta} \frac{1}{\tilde{\Omega}\tilde{D}}\left(\tilde{v}_n(1 - \phi^{(0)}) + \frac{\tilde{\xi}_1\tilde{v}_n}{2}(\phi_{\zeta}^{(0)})^2\right)d\eta\,. \tag{5.101}$$

The discontinuity of the adatom concentration field across the boundary Γ_{ij} now reads:

$$C_+^{(1)} - C_-^{(1)} = \frac{\tilde{\xi}_1\tilde{v}_n}{12}\,. \tag{5.102}$$

Evaluation at the front and at the back of each boundary Γ_{ij} and inserting into the full expansion for C up to the first order results in:

$$(C_+ - C_{\text{eq}}) \frac{12\mathcal{X}}{\tilde{\xi}_1 l} = \frac{\partial C_+}{\partial n} \,, \tag{5.103}$$

and

$$(C_- - C_{\text{eq}}) \frac{12\mathcal{X}}{\tilde{\xi}_1 l} = -\frac{\partial C_-}{\partial n} \,, \tag{5.104}$$

respectively. Thus the equivalents of (5.69) and (5.70) are recovered. This validates that the model equations (5.80) and (5.81) are an equivalent formulation of the full set of equations constituting the sharp interface formulation (5.68)–(5.71).

5.3 A Generalized Approach Towards the Asymptotic Analysis of Diffuse Interface Models

Whereas in the previous section the thin interface analysis arose as an extension of the sharp interface analysis, in this section I discuss a generalized approach towards matching, which directly incorporates a notion of the diffuse interface. It was proposed by Elder, Grant, Provatas and Kosterlitz in [102] and works for a specific class of moving boundary problems only. These include order/disorder transitions, spinodal decomposition, Ostwald ripening, dendritic growth and eutectic alloy solidification. For this class of phenomena the projection operator method can be employed to extract the sharp interface limit from diffuse interface models with interfaces that are diffuse on a length scale ξ. In particular, phase-field equations are mapped onto sharp interface equations in the limits $\xi\kappa \ll 1$ and $\xi v/D \ll 1$, where κ and v are the interface curvature and velocity, respectively. D is the diffusion constant in the bulk. The approach follows the projection operator method of Kawasaki and Ohta [173, 174, 175, 176, 225]. It is generally applicable in the sense that the free energy functional $\mathcal{F}$ does not need to be specified. This implies that the free energy density can be chosen freely, such that it guarantees strict relaxational behavior of the thermodynamic potential and thus ensures thermodynamic consistency inherently.

5.3.1 Role of the Governing Thermodynamic Potential

Following [102] I will consider two fields, a non-conserved phase-field ϕ and a conserved field c. The phase-field distinguishes between the two phases, e.g. liquid phase and solid phase. The c field can be understood as a concentration. The free energy functional describing the system can be written as

$$\mathcal{F}\{\varPhi, c\} = \int d\mathbf{r} \left[\frac{1}{2} K_\phi |\nabla\phi|^2 + \frac{1}{2} K_c |\nabla c|^2 + f(\phi, c) \right] , \tag{5.105}$$

where $f(\phi, c)$ is the local bulk free energy density. The gradient terms are related to the phase-separating interface itself. The dynamics of these fields are described by an equation of motion for the non-conserved variable ϕ,

$$\frac{\partial \phi}{\partial t} = -\Gamma_\phi \frac{\delta \mathcal{F}}{\delta \phi} = -\Gamma_\phi \left[-K_\phi \nabla^2 \phi + \frac{\partial f}{\partial \phi} \right] , \qquad (5.106)$$

as well as an equation of motion for the conserved variable c, i.e. the concentration:

$$\frac{\partial c}{\partial t} = \Gamma_c \nabla^2 \frac{\delta \mathcal{F}}{\delta c} = \Gamma_c \nabla^2 \left[-K_c \nabla^2 c + \frac{\partial f}{\partial c} \right] . \qquad (5.107)$$

Here $\delta \mathcal{F} \{ \Phi, c \}/\delta c = \mu$ refers to the chemical potential. Γ_c and Γ_ϕ are the respective mobilities (compare to (4.42) of Sect. 4.1).

A generic bulk free energy density $f(c, \phi)$ can give rise to a variety of equilibrium as well as a variety of metastable states. To illustrate this, I will consider the following bulk free energy

$$f(\phi, c) = u \frac{T}{T_M} \left[c \ln c + (1 - c) \ln (1 - c) \right]$$
$$+ \left(\alpha \Delta T - \beta (c - \frac{1}{2})^2 \right) \Phi(\phi) - \frac{1}{2} \phi^2 + \frac{1}{4} \phi^4 . \qquad (5.108)$$

Here $\Phi(\phi) \equiv 2\phi - 4\phi^3/3 + 2\phi^5/5$, $\Delta T \equiv (T - T_M)/T_M$, where T_M denotes the melting temperature. The parameters α, β and u are phenomenological parameters and have to be determined by calibration to experimental phase diagrams. If these parameters are chosen as $\alpha = \beta = 1.0$ and $u = 0.6$, a phase diagram as displayed in Fig. 5.3 emerges. A different type of phase diagram arising from parameters $\alpha = \beta = 1.0$ and $u = 0.45$ is pictured in Fig. 5.4. Obviously the choice of parameters determines differently shaped domains of pure phases (i.e. liquid phase L and solid phase S) as well as coexistence regimes $S + L$ (solid/liquid) and $S + S$ (solid/solid) (compare to the phase diagram explained in detail in Sect. 3.1, depicted by Fig. 3.1). Within the current section the focus is on the dynamics of phase-separating interfaces. The dynamics of a system can result in transitions from one point of the phase diagram to another. Different such transitions are shown in Fig. 5.3 and Fig. 5.4. Each of them denotes a different kind of growth phenomenon. They are termed *quenches*. More precisely a quench is defined as a rapid change in temperature taking a system from one domain within the phase diagram to another. It is often considered instantaneous in theoretical modeling. The generalized calculations of the following subsection are carried out including all the possible quenches illustrated in Fig. 5.3 and Fig. 5.4. As stated above, this does not require to specify the explicit form of the bulk free energy density f. It is simply assumed that f is chosen in a way such that all the phases of interest are well defined.

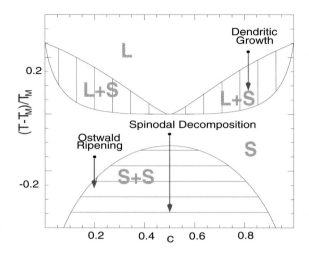

Fig. 5.3. Mean-field phase diagram obtained from a bulk free energy given by (5.108) with parameters $\alpha = \beta = 1.0$ and $u = 0.6$. In this figure the regions containing vertical and horizontal lines are liquid/solid $(L + S)$ and solid/solid $(S+S)$ coexistence regions, respectively. The figure follows, just as further figures in this section, [102]. Here liquid/solid and solid/solid coexistence regimes are disjunct. Decreasing the parameter u within the energy density f lowers the critical point T_c of possible solid/solid coexistence, so that for smaller values of u the two coexistence regimes will merge. Such a situation is depicted in Fig. 5.4. A phase diagram with merging coexistence regimes gives rise to a qualitatively different *quench* (see text), which refers to the growth phenomenon of eutectic solidification.

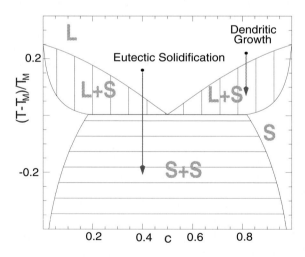

Fig. 5.4. Same as the previous figure apart from different parameters for the bulk free energy given by $\alpha = \beta = 1.0$ and $u = 0.45$.

5.3.2 The Expansion Procedure in Detail

The goal of this section is to derive the sharp interface formulation for the dynamics of a system described by a free energy functional $\mathcal{F}$ as given in (5.105) and governed by (5.106) and (5.107) via the generalized asymptotic analysis. It proceeds by expanding around a planar equilibrium interface. The interface is out of equilibrium as soon as it is either gently curved or one of the two bulk phases is metastable. In the former case, a curvature is assumed "gentle", if the radius of curvature $1/\kappa$ is large compared to the interface width or correlation length. This yields one small *dimensionless* expansion parameter $\kappa\xi$. In the latter case, the difference between the free energy of the stable and metastable phases causes the interface to propagate into the metastable one. If the free energy difference is small the propagation velocity v is small, as well. In this context a velocity can be interpreted to be "small", if the interface moves so slowly that a steady state diffusion field is allowed to form in front of the interface. This implies that the time for the diffusion field to relax during interface motion of a distance ξ should be much smaller than the time ξ/v the interface itself requires to move that distance. Since the diffusion time is given by $\tau = \xi^2/D$, this leads to a second small *dimensionless* parameter $\xi v/D$. The latter is known as the interfacial Peclet number. In the following analysis the interface equations will be obtained to lowest order in both small parameters. Technically the expansion to lowest order in these two parameters can be achieved by regarding them to be of the same order in the expansion. In the calculations to follow both parameters will be considered $\mathcal{O}(\varepsilon)$ with $\varepsilon \ll 1$.

The succeeding expansion exploits the fact that the fields behave qualitatively different close and far from the interface. In the region close to the interface, they vary rapidly over distances $\mathcal{O}(\xi)$, while far from the interface they vary on a scale $\mathcal{O}(\xi/\varepsilon)$. If a length scale ζ exists such that $1 \ll \zeta/\xi \ll 1/\varepsilon$, distinct "inner" and "outer" regions can be defined. Those are depicted in Fig. 5.5. For the analysis to follow it is appropriate to expand in both inner and outer regions and match the solutions order by order at the length scale ζ. In this sense the expansion follows the ideas described previously in Sect. 5.1.2. It differs from the example provided to illustrate the "usual" way to carry out matching in the *thin interface limit* with fully specified potentials and respective thin interface analysis model equations in Sect. 5.2 in three points:

1. The generalized analysis is carried out to lowest order in *two* dimensionless parameters, i.e. $\kappa\xi$ and the interfacial Peclet number $Pe_{\mathrm{Int}} = \xi v/D$. Technically this is possible by considering them both of $\mathcal{O}(\varepsilon)$ (see above).
2. The expansion involves usage of the Greens function formalism and further the projection operator method as discussed in the introductory part of this section. This allows us to remain on a general level.

3. An additional solvability condition is obtained by employing the definition of the *Gibbs surface*. It is required to carry out the matching – again
   remaining on a general level.

Following [102] all of the above is featured in the succeeding paragraphs.

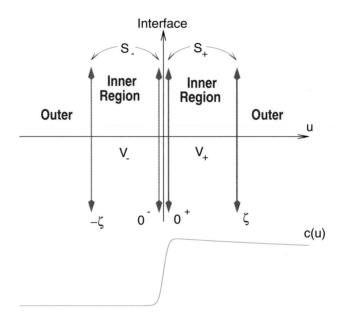

Fig. 5.5. Illustration of inner and outer regions defined in the context of the
expansion procedure described within this section.

Inner Expansion I start by discussing the inner expansion, which applies
to a region defined by $-\zeta < u < \zeta$. Here u is a coordinate normal to the
interface and $1 \ll \zeta/\xi \ll 1/\varepsilon$. The aim is to obtain asymptotic expansions
for the solutions of the evolution equations (5.106) and (5.107) valid in this
inner region. The latter can be written in a compact form as

$$\frac{1}{\Gamma_c}\frac{\partial c}{\partial t} = \nabla^2 \delta\mu \qquad (5.109)$$

where

$$\delta\mu = \mu(\mathbf{r}) - \mu_{\mathrm{eq}} = -K_c\nabla^2 c + \frac{\partial f}{\partial c} - \mu_{\mathrm{eq}} . \qquad (5.110)$$

The first step is to partition the system into two regions V_+ and V_-
bounded by surfaces S_+ and S_-, respectively. The region V_+ is defined by
$0 < u < \zeta$, V_- by $-\zeta < u < 0$. The position of the interface between two

bulk regions is defined as $u(\mathbf{r}, t) = 0$. This definition is a formal one. To get an impression one may picture the surface $u(\mathbf{r}, t) = 0$ as the surface near which the fields c and ϕ vary rapidly over a distance $\mathcal{O}(\xi)$. However the exact position remains unspecified at this point. Moreover it is useful to define Greens functions $G^{\pm}(\mathbf{r}, \mathbf{r}')$ in the regions $-\zeta < u < 0$ (G^{-}) and $0 < u < \zeta$ (G^{+}), respectively, obeying

$$\nabla'^2 G^{\pm}(\mathbf{r}, \mathbf{r}') = \delta(\mathbf{r} - \mathbf{r}') . \tag{5.111}$$

These Greens functions satisfy the following boundary conditions:

- $G(\mathbf{r}, \mathbf{r}') = 0$ at $u = 0$ and $u' = 0$,
- $\partial G(\mathbf{r}, \mathbf{r}')/\partial u = 0$ at $u = \pm\zeta$,
- $\partial G(\mathbf{r}, \mathbf{r}')/\partial u' = 0$ at $u' = \pm\zeta$,
- periodic at $u = \pm\infty$ and $u' = \pm\infty$.

Note that *both* $\mathbf{r}$ and $\mathbf{r}'$ lie in the same region V_+ or V_-, respectively.
    Multiplying (5.109) by $G^{\pm}$ and integrating over $\mathbf{r}' \in V_{\pm}$ yields

$$\delta\mu(\mathbf{r}) = \int_{V_{\pm}} d\mathbf{r}' \frac{G^{\pm}(\mathbf{r}, \mathbf{r}')}{\Gamma_c} \frac{\partial c'}{\partial t}$$
$$+ \oint_{S_{\pm}} d\mathbf{S}' \cdot \left(\delta\mu' \nabla' G^{\pm} - G^{\pm} \nabla' \delta\mu' \right) , \tag{5.112}$$

where $\delta\mu' = \delta\mu(\mathbf{r}')$ is defined as in (5.110). $\int_{V_+} d\mathbf{r}'$ denotes integration over V_+ (i.e. $0 < u(\mathbf{r}') < \zeta$). Moreover, $\int_{V_-} d\mathbf{r}'$ refers to integration over V_-. In analogy, $\oint_{S_+} d\mathbf{S}'$ denotes integration over the boundary S_+ enclosing V_+. Similarly $\oint_{S_-} d\mathbf{S}'$ denotes integration over the boundary S_- enclosing V_-. Multiplying (5.112) by $\partial c_0^{in}/\partial u$ and integrating over $-\zeta < u < \zeta$ yields

$$\mathcal{B} + \mathcal{S} \equiv \int_{-\zeta}^{\zeta} du \frac{\partial c_0^{in}}{\partial u} \delta\mu$$
$$= \int_{-\zeta}^{\zeta} du \frac{\partial c_0^{in}}{\partial u} \left(\frac{\partial f}{\partial c} - K_c \nabla^2 c - \mu_{eq} \right) , \tag{5.113}$$

where $\mathcal{B} = \mathcal{B}^+ + \mathcal{B}^-$ and $\mathcal{S} = \mathcal{S}^+ + \mathcal{S}^-$ with

$$\mathcal{B}^{\pm} = \pm \frac{1}{\Gamma_c} \int_0^{\pm\zeta} du \frac{\partial c_0^{in}}{\partial u} \int_{V_{\pm}} d\mathbf{r}' G^{\pm}(\mathbf{r}, \mathbf{r}') \frac{\partial c'}{\partial t} , \tag{5.114}$$

$$\mathcal{S}^{\pm} = \pm \int_0^{\pm\zeta} du \frac{\partial c_0^{in}}{\partial u} \oint_{S_{\pm}} d\mathbf{S}' \cdot \left(\delta\mu' \nabla' G^{\pm} - G^{\pm} \nabla' \delta\mu' \right) . \tag{5.115}$$

An analogous formula for ϕ can be derived by multiplying (5.106) by $\partial\phi_0^{in}/\partial u$ and integrating over $-\zeta < u < +\zeta$ to obtain

$$\frac{1}{\Gamma_\phi} \int_{-\zeta}^{\zeta} du \frac{\partial \phi_0^{\text{in}}}{\partial u} \frac{\partial \phi}{\partial t} = + \int_{-\zeta}^{\zeta} du \frac{\partial \phi_0^{\text{in}}}{\partial u} \left(K_\phi \nabla^2 \phi - \frac{\partial f}{\partial \phi} \right) . \tag{5.116}$$

Each term in the above equations can systematically be expanded in powers of ε. In this section, attention is restricted to the terms $\mathcal{O}(\varepsilon)$, which will prove to be sufficient to obtain the relevant sharp interface conditions.

To facilitate the expansion, $c(\mathbf{r}, t)$, $\phi(\mathbf{r}, t)$ and the chemical potential $\mu(\mathbf{r}, t)$ are expressed by the help of a power series in ε as follows:

$$\begin{aligned} c(\mathbf{r}, t) &= c_0^{\text{in}}(u(\mathbf{r})) + \varepsilon \delta c_1^{\text{in}} + \varepsilon^2 \delta c_2^{\text{in}} + \cdots \\ \phi(\mathbf{r}, t) &= \phi_0^{\text{in}}(u(\mathbf{r})) + \varepsilon \delta \phi_1^{\text{in}} + \varepsilon^2 \delta \phi_2^{\text{in}} + \cdots \\ \mu(\mathbf{r}, t) &= \mu_0^{\text{in}}(u(\mathbf{r})) + \varepsilon \delta \mu_1^{\text{in}} + \varepsilon^2 \delta \mu_2^{\text{in}} + \cdots , \end{aligned} \tag{5.117}$$

where the superscript "in" refers to the *inner* solution. To expand the Laplacian in powers of ε, it is useful to introduce a curvilinear coordinate system with one coordinate u along the local normal to the interface and $(d-1)$ coordinates $\mathbf{s}$ perpendicular to u and tangential to the interface. For simplicity a two-dimensional system is considered, where s is the scalar arc length. Note that within (5.117) the $\mathcal{O}(\varepsilon^0)$ terms c_0^{in} and ϕ_0^{in} are the equilibrium planar interface solutions.

At this stage I come back to the point, that the exact position of the interface has not yet been specified. Its choice is somewhat flexible to within a distance ξ. It is one of the key elements of this generalized analysis to choose it in accordance to the definition of the Gibbs surface, so that the excess surface concentration is equal on both sides of the interface. This results in an additional solvability condition, which ensures that the chemical potential is continuous across the interface.

The transformation from Cartesian to curvilinear coordinates (see Appendix A) results in the formal expansion

$$\xi^2 \nabla^2 = \mathcal{L}_0 + \epsilon \mathcal{L}_1 + \epsilon^2 \mathcal{L}_2 + \cdots , \tag{5.118}$$

where the specific form of $\mathcal{L}_n$ depends on the expansion. In the inner region, derivatives of the fields with respect to u are much larger than derivatives with respect to s, which are zero, if the curvature and the interfacial Peclet number vanish. This is taken into account by introducing the dimensionless variables $\bar{u}$ and $\bar{s}$. These are of $\mathcal{O}(\varepsilon^0)$ and given by $u = \xi \bar{u}$ and $s = \xi \bar{s}/\varepsilon$. As proven in Appendix A, this scaling results in:

$$\begin{aligned} \mathcal{L}_0 &= \partial^2 / \partial \bar{u}^2 \\ \mathcal{L}_1 &= \bar{\kappa} \partial / \partial \bar{u} \\ \mathcal{L}_2 &= \partial^2 / \partial \bar{s}^2 - \bar{\kappa}^2 \bar{u} \partial / \partial \bar{u} , \end{aligned}$$

where the dimensionless curvature $\bar{\kappa} \equiv \xi \kappa / \varepsilon$ is of order unity.

A final step is to expand the time derivatives within (5.114) and (5.116) in ϵ. For these calculations, it is convenient to choose a frame of reference, in which the interface position is stationary, such that

$$\frac{\partial}{\partial t}\Big|_{\mathbf{r}} = \frac{\partial}{\partial t}\Big|_{(u,s)} - \mathbf{v} \cdot \nabla \, . \tag{5.119}$$

Here $\mathbf{v}$ is the interface velocity which has components normal and tangential to the interface. The time derivative on the right hand side corresponds to a relaxational dynamics independent of interfacial motion. If this operator acts on the fields c and ϕ, the tangential component and time derivative are of order ε^3. Thus they can be dropped[4]. To $\mathcal{O}(\varepsilon)$, $\partial/\partial t|_{\mathbf{r}}$ becomes:

$$\frac{\partial}{\partial t}\Big|_{\mathbf{r}} = -\varepsilon \frac{v_1}{\tau}\frac{\partial}{\partial \bar{u}} + \mathcal{O}(\varepsilon)^2 + \cdots \, , \tag{5.120}$$

where the normal velocity has been expanded in a power series in ε, i.e.:

$$v_{\mathrm{n}} \equiv -\frac{\partial u}{\partial t} \equiv \frac{\xi}{\tau}\sum_{m=1}^{\infty}\varepsilon^m v_m \, . \tag{5.121}$$

Using these expansions and further expanding f around c_0^{in} and ϕ_0^{in}, the right hand sides of (5.113) and (5.116) become

$$\int_{-\zeta}^{\zeta} du \frac{\partial c_0^{\mathrm{in}}}{\partial u}\left[\mu_0 + \bar{K}_c \xi^2 \nabla^2 c - \frac{\partial f}{\partial c}\right] = \int_{-\bar{\zeta}}^{\bar{\zeta}} d\bar{u}\frac{\partial c_0^{\mathrm{in}}}{\partial \bar{u}}\left[\left(\mu_0 + \bar{K}_c \mathcal{L}_0 c_0^{\mathrm{in}} - f_i^{(1,0)}\right)\right.$$
$$\left. + \epsilon\left(\bar{K}_c(\mathcal{L}_1 c_0^{\mathrm{in}} + \mathcal{L}_0 \delta c_1^{\mathrm{in}}) - \delta c_1^{\mathrm{in}} f_i^{(2,0)} - \delta\phi_1^{\mathrm{in}} f_i^{(1,1)}\right) + \mathcal{O}(\epsilon^2)\right] \tag{5.122}$$

and

$$\int_{-\zeta}^{\zeta} du \frac{\partial \phi_0^{\mathrm{in}}}{\partial u}\left[\bar{K}_\phi \xi^2 \nabla^2 \phi - \frac{\partial f}{\partial \phi}\right]$$
$$= \int_{-\bar{\zeta}}^{\bar{\zeta}} d\bar{u}\frac{\partial \phi_0^{\mathrm{in}}}{\partial \bar{u}}\left[\left(\bar{K}_\phi \mathcal{L}_0 \phi_0^{\mathrm{in}} - f_i^{(0,1)}\right) + \epsilon\left(\bar{K}_\phi(\mathcal{L}_1 \phi_0^{\mathrm{in}}\right.\right.$$
$$\left.\left. + \mathcal{L}_0 \delta\phi_1^{\mathrm{in}}) - \delta\phi_1^{\mathrm{in}} f_i^{(0,2)} - \delta c_1^{\mathrm{in}} f_i^{(1,1)}\right) + \mathcal{O}(\epsilon^2)\right] \, . \tag{5.123}$$

Here $f_i^{(n,m)} \equiv \partial^{n+m} f/\partial c^n \partial \phi^m|_{\phi_0^{\mathrm{in}}, c_0^{\mathrm{in}}}$, $\bar{K}_c \equiv K_c/\xi^2$, $\bar{K}_\phi \equiv K_\phi/\xi^2$ and $\bar{\zeta} = \zeta/\xi$. Terms of $\mathcal{O}(\epsilon^0)$ vanish by construction. For later use, it is convenient to perform partial integrations on combinations of terms:

---

[4] In two dimensions the tangential component of the time derivative can be expressed as $v_T \partial/\partial s$, where v_T accounts for flow of concentration (or ϕ) along the interface. v_T must be $\mathcal{O}(\varepsilon)$ since it is zero for $\varepsilon = 0$. Furthermore in the inner region s is scaled with ε such that $v_T \partial/\partial s \sim \mathcal{O}(\varepsilon^2)$. If this term operates on c or ϕ, it becomes $\mathcal{O}(\varepsilon^3)$, since both c and ϕ are independent of s. The time derivative in the moving frame $\partial/\partial t|_{(u,s)}$ is at least of order $\mathcal{O}(\varepsilon^3)$. This term accounts for the relaxation of fluctuations around the steady state profile. For the concentration field fluctuations relax as $t \sim (\text{length})^2$, which implies $\partial/\partial t|_{(u,s)} \sim \mathcal{O}(\varepsilon^2)$. Just as the tangential component this term becomes $\mathcal{O}(\varepsilon^3)$ when operating on c. Moreover, ϕ fluctuations relax exponentially fast, thus rendering the term negligible.

$$\int_{-\bar{\zeta}}^{\bar{\zeta}} \mathrm{d}\,\bar{u}\,\frac{\partial c_0^{\mathrm{in}}}{\partial \bar{u}}\left(\bar{K}_c \frac{\partial^2}{\partial \bar{u}^2} - f_i^{(2,0)}\right)\delta c_1^{\mathrm{in}} = \int_{-\bar{\zeta}}^{\bar{\zeta}} \mathrm{d}\bar{u}\,\delta c_1^{\mathrm{in}}\left(\bar{K}_c \frac{\partial^2}{\partial \bar{u}^2} - f_i^{(2,0)}\right)\frac{\partial c_0^{\mathrm{in}}}{\partial \bar{u}}$$

$$= \int_{-\bar{\zeta}}^{\bar{\zeta}} \mathrm{d}\bar{u}\,\delta c_1^{\mathrm{in}} f_i^{(1,1)}\frac{\partial \phi_0^{\mathrm{in}}}{\partial \bar{u}} \tag{5.124}$$

and

$$\int_{-\bar{\zeta}}^{\bar{\zeta}} \mathrm{d}\bar{u}\,\frac{\partial \phi_0^{\mathrm{in}}}{\partial \bar{u}}\left(\bar{K}_\phi \frac{\partial^2}{\partial \bar{u}^2} - f_i^{(0,2)}\right)\delta\phi_1^{\mathrm{in}} = \int_{-\bar{\zeta}}^{\bar{\zeta}} \mathrm{d}\bar{u}\,\delta\phi_1^{\mathrm{in}}\left(\bar{K}_\phi \frac{\partial^2}{\partial \bar{u}^2} - f_i^{(0,2)}\right)\frac{\partial \phi_0^{\mathrm{in}}}{\partial \bar{u}}$$

$$= \int_{-\bar{\zeta}}^{\bar{\zeta}} \mathrm{d}\bar{u}\,\delta\phi_1^{\mathrm{in}} f_i^{(1,1)}\frac{\partial c_0^{\mathrm{in}}}{\partial \bar{u}}\,, \tag{5.125}$$

since derivatives of c_0^{in} and ϕ_0^{in} vanish at $\bar{u}=\pm\bar{\zeta}$ in the limit $\bar{\zeta}=\zeta/\xi \gg 1$. Here the final equations (5.124) and (5.125) were obtained using the relationships

$$\frac{\partial}{\partial \bar{u}}\left(\frac{\partial f(c,\phi)}{\partial c}\right) = \frac{\partial^2 f(c,\phi)}{\partial c^2}\frac{\partial c}{\partial \bar{u}} + \frac{\partial^2 f(c,\phi)}{\partial c\partial\phi}\frac{\partial \phi}{\partial \bar{u}} = f^{(2,0)}\frac{\partial c}{\partial \bar{u}} + f^{(1,1)}\frac{\partial \phi}{\partial \bar{u}}$$

and

$$\frac{\partial}{\partial \bar{u}}\left(\frac{\partial f(c,\phi)}{\partial \phi}\right) = \frac{\partial^2 f(c,\phi)}{\partial \phi^2}\frac{\partial \phi}{\partial \bar{u}} + \frac{\partial^2 f(c,\phi)}{\partial c\partial\phi}\frac{\partial c}{\partial \bar{u}} = f^{(0,2)}\frac{\partial \phi}{\partial \bar{u}} + f^{(1,1)}\frac{\partial c}{\partial \bar{u}}\,.$$

To complete the calculation, the left-hand sides of (5.113) and (5.116) are expanded to lowest order in ϵ. The expansion for ϕ in (5.116) is straightforward

$$\frac{1}{\Gamma_\phi}\int_{-\zeta}^{\zeta} \mathrm{d}u\,\frac{\partial \phi_0^{\mathrm{in}}}{\partial u}\frac{\partial \phi}{\partial t} = -\frac{\epsilon}{\Gamma_\phi \tau}\int_{-\bar{\zeta}}^{\bar{\zeta}} \mathrm{d}\bar{u}\,\frac{\partial \phi_0^{\mathrm{in}}}{\partial \bar{u}}v_1\frac{\partial \phi_0^{\mathrm{in}}}{\partial \bar{u}}$$

$$= -\varepsilon\frac{v_1\xi}{\tau}\frac{\sigma_\phi}{K_\phi\Gamma_\phi} + \mathcal{O}(\varepsilon^2)\,, \tag{5.126}$$

where

$$\sigma_\phi \equiv K_\phi \int_{-\zeta}^{\zeta} \mathrm{d}u\left(\frac{\partial \phi_0^{\mathrm{in}}}{\partial u}\right)^2\,. \tag{5.127}$$

The equivalent expansion for c is more complicated. The algebra is given in Appendix B. Formally the results of these calculations can be expressed as

$$\mathcal{B} + \mathcal{S} = \epsilon(\mathcal{B}_1/\tau + \mathcal{S}_1) + \epsilon^2(\mathcal{B}_2/\tau + \mathcal{S}_2) + \cdots\,, \tag{5.128}$$

where $\mathcal{B}_n$ and $\mathcal{S}_n$ are given in Appendix B.

Combining all of the above yields the following two equations to $\mathcal{O}(\epsilon)$:

$$\frac{v_1\xi}{\tau}\frac{\sigma_\phi}{\Gamma_\phi K_\phi} = -\frac{\bar{\kappa}\sigma_\phi}{\xi} - A_1 \tag{5.129}$$

and

$$\Delta c\,\delta\mu_1^{\rm in}(0,s) = -\frac{\sigma_c\bar{\kappa}}{\xi} + A_1 - \frac{v_1(s)\xi^2}{\tau\Gamma_c}\int_{-\bar{\zeta}}^{\bar{\zeta}}{\rm d}\bar{u}[c_0^{\rm out}(\bar{u}) - c_0^{\rm in}(\bar{u})]^2$$
$$- \frac{\partial\delta\mu_1^{\rm in}}{\partial\bar{u}}\bigg|_{-\bar{\zeta}}\int_{-\bar{\zeta}}^{0}{\rm d}\bar{u}[c_0^{\rm out}(\bar{u}) - c_0^{\rm in}(\bar{u})]$$
$$- \frac{\partial\delta\mu_1^{\rm in}}{\partial\bar{u}}\bigg|_{\bar{\zeta}}\int_{0}^{\bar{\zeta}}{\rm d}\bar{u}[c_0^{\rm out}(\bar{u}) - c_0^{\rm in}(\bar{u})]\;, \tag{5.130}$$

where $\Delta c \equiv c_0^{\rm in}(\bar{\zeta}) - c_0^{\rm in}(-\bar{\zeta})$ is the miscibility gap, and

$$c_0^{\rm out}(\bar{u}) \equiv \begin{cases} c_0^{\rm in}(-\bar{\zeta}) & \bar{u} < 0 \\ c_0^{\rm in}(+\bar{\zeta}) & \bar{u} > 0 \end{cases}, \tag{5.131}$$

$$\sigma_c \equiv K_c\int_{-\zeta}^{\zeta}{\rm d}u\left(\frac{\partial c_0^{\rm in}}{\partial u}\right)^2, \tag{5.132}$$

$$A_1 = \int_{-\bar{\zeta}}^{\bar{\zeta}}{\rm d}\bar{u}\left(\delta\phi_1^{\rm in}\frac{\partial c_0^{\rm in}}{\partial\bar{u}} - \delta c_1^{\rm in}\frac{\partial\phi_0^{\rm in}}{\partial\bar{u}}\right)f_i^{(1,1)}. \tag{5.133}$$

Equation (5.130) yields the chemical potential μ of the *inner* solution at the interface (i.e., at $\bar{u} = 0$). This must be matched to the *outer* solution at $\bar{u} = \pm\bar{\zeta}$. An expression for $\delta\mu_1^{\rm in}(\pm\bar{\zeta})$ can be obtained by expanding (5.109) to $\mathcal{O}(\epsilon)$:

$$v_1\frac{\partial c_0^{\rm in}}{\partial\bar{u}} = -\frac{\Gamma_c\tau}{\xi^2}\mathcal{L}_0\,\delta\mu_1^{\rm in}. \tag{5.134}$$

Integrating (5.134) twice - first from $\bar{u}$ to $\bar{\zeta}$, then from 0 to $\bar{\zeta}$ - yields

$$\delta\mu_1^{\rm in}(\bar{\zeta},s) = \delta\mu_1^{\rm in}(0,s) + \bar{\zeta}\frac{\partial\delta\mu_1^{\rm in}}{\partial\bar{u}}\bigg|_{\bar{\zeta}} + \frac{v_1\xi^2}{\tau\Gamma_c}\int_{0}^{\bar{\zeta}}{\rm d}\bar{u}[c_0^{\rm out}(\bar{u}) - c_0^{\rm in}(\bar{u})]. \tag{5.135}$$

A similar procedure has to be applied to $\delta\mu_1^{\rm in}(-\bar{\zeta})$. From (5.130) and (5.135) one obtains

$$\Delta\,c\delta\mu_1^{\rm in}(\pm\bar{\zeta},s) =$$
$$-\sigma_c\kappa/\epsilon + A_1 \pm \Delta c\bar{\zeta}\frac{\partial\delta\mu_1^{\rm in}}{\partial\bar{u}}\bigg|_{\pm\bar{\zeta}} - \frac{v_1\xi^2}{\tau\Gamma_c}\int_{-\bar{\zeta}}^{\bar{\zeta}}{\rm d}\bar{u}[c_0^{\rm out}(\bar{u}) - c_0^{\rm in}(\bar{u})]^2$$
$$- \frac{\partial\delta\mu_1^{\rm in}}{\partial\bar{u}}\bigg|_{-\bar{\zeta}}\int_{-\bar{\zeta}}^{0}{\rm d}\bar{u}[c_0^{\rm out}(\bar{u}) - c_0^{\rm in}(\bar{u})] - \frac{\partial\delta\mu_1^{\rm in}}{\partial\bar{u}}\bigg|_{\bar{\zeta}}\int_{0}^{\bar{\zeta}}{\rm d}\bar{u}[c_0^{\rm out}(\bar{u}) - c_0^{\rm in}(\bar{u})]$$
$$+ \Delta c\frac{v_1\xi^2}{\tau\Gamma_c}\int_{0}^{\pm\bar{\zeta}}{\rm d}\bar{u}[c_0^{\rm out}(\bar{u}) - c_0^{\rm in}(\bar{u})]. \tag{5.136}$$

The integrals in (5.136) can be rewritten in a more suitable form by noting that $\bar\zeta \gg 1$ in the inner region, so that $c_0^{\rm in}(\pm\bar u) = c_0^{\rm in}(\pm\infty)$ for $|\bar u| \geq \bar\zeta$. Thus (5.136) results in

$$
\begin{aligned}
\Delta\, c\delta\mu_1^{\rm in}(\pm\bar\zeta, \bar s) =\\
-\frac{\sigma_c\bar\kappa}{\xi} + A_1 \pm \Delta c\bar\zeta\frac{\partial\delta\mu_1^{\rm in}}{\partial\bar u}\Big|_{\pm\bar\zeta} - \frac{v_1(\bar s)\xi^2}{\tau\Gamma_c}\int_{-\infty}^{\infty} \mathrm{d}\bar u[c_0^{\rm in}(\bar u) - c_0^{\rm out}(\bar u)]^2\\
-\frac{\partial\delta\mu_1^{\rm in}}{\partial\bar u}\Big|_{-\bar\zeta}\int_{-\infty}^{0} \mathrm{d}\bar u[c_0^{\rm out}(\bar u) - c_0^{\rm in}(\bar u)] - \frac{\partial\delta\mu_1^{\rm in}}{\partial\bar u}\Big|_{\bar\zeta}\int_{0}^{\infty} \mathrm{d}\bar u[c_0^{\rm out}(\bar u) - c_0^{\rm in}(\bar u)]\\
+\Delta c\frac{v_1(\bar s)\xi^2}{\tau\Gamma_c}\int_{0}^{\pm\infty} \mathrm{d}\bar u[c_0^{\rm out}(\bar u) - c_0^{\rm in}(\bar u)]\,.
\end{aligned}
\tag{5.137}
$$

At this point the expression for v_1 is still lacking. It is obtained by integrating (5.110) over $-\bar\zeta < \bar u < \bar\zeta$

$$
v_1 = -\frac{\tau\Gamma_c}{\Delta c\xi^2}\left(\frac{\partial\delta\mu_1^{\rm in}}{\partial\bar u}\Big|_{\bar\zeta} - \frac{\partial\delta\mu_1^{\rm in}}{\partial\bar u}\Big|_{-\bar\zeta}\right).
\tag{5.138}
$$

This solution for $\delta\mu_1^{\rm in}(\pm\bar\zeta)$ has to be matched to the solution of the expansion within the *outer* region, to which I will turn next.

Outer Expansion Far from the interface, the fields ϕ and c vary slowly in space and are close to the bulk equilibrium values $\phi_{\rm eq}$ and $c_{\rm eq}$. Variations of the fields in the bulk regions far from the interface take place on length scales $\mathcal{O}(\xi/\epsilon)$ in all directions. Therefore suitable dimensionless space and time coordinates are $(\tilde u, \tilde s, \tilde t) \equiv (\epsilon u/\xi, \epsilon s/\xi, \epsilon^2 t/\tau)$.

Expanding $\phi(\mathbf r)$ around the bulk equilibrium solution $\phi(\mathbf r) = \delta\phi^{\rm out}(\mathbf r) + \phi_{\rm eq}$ yields

$$
\begin{aligned}
\frac{\partial\delta\phi^{\rm out}}{\partial t} = \Gamma_\phi\Bigg(K_\phi\nabla^2\delta\phi^{\rm out} - \frac{\partial f}{\partial\phi}\Big|_{\rm eq} - \frac{\partial^2 f}{\partial\phi^2}\Big|_{\rm eq}\delta\phi^{\rm out}\\
-\frac{1}{2!}\frac{\partial^3 f}{\partial\phi^3}\Big|_{\rm eq}(\delta\phi_0^{\rm out})^2 - \frac{1}{3!}\frac{\partial^4 f}{\partial\phi^4}\Big|_{\rm eq}(\delta\phi^{\rm out})^3 - \cdots\Bigg).
\end{aligned}
\tag{5.139}
$$

By definition, $(\partial f/\partial\phi)_{\rm eq} = 0$. Since $\delta\phi^{\rm out} = 0$ in the limit $\epsilon \to 0$, (5.139) is linear at $\mathcal{O}(\epsilon)$. Furthermore, $(\partial^2 f/\partial\phi^2)_{\rm eq} > 0$, so that $\delta\phi_0^{\rm out}$ vanishes exponentially with time for all wavelengths. Thus, $\delta\phi_0^{\rm out}$ is trivial in the outer region and can be ignored. It is convenient to expand $c^{\rm out}$ and $\mu^{\rm out}$ in the outer region as

$$
\begin{aligned}
c^{\rm out}(\mathbf r) &= c_0^{\rm out} + \epsilon\delta c_1^{\rm out} + \cdots\\
\mu^{\rm out}(\mathbf r) &= \mu_0^{\rm out} + \epsilon\delta\mu_1^{\rm out} + \cdots,
\end{aligned}
\tag{5.140}
$$

where $c_0^{\rm out}$ is given by (5.131). At $\mathcal{O}(\epsilon^3)$ in dimensionless variables:

$$
\frac{\partial\delta c_1^{\rm out}}{\partial\tilde t} = \frac{\tau\Gamma_c}{\xi^2}\tilde\nabla^2\delta\mu_1^{\rm out},
\tag{5.141}
$$

where $\tilde{\nabla} \equiv (\xi/\epsilon)\nabla$ is the scaled dimensionless derivative suitable for the outer region. Equation (5.140) is evaluated in a laboratory frame. It simplifies to a linear diffusion equation for the chemical potential inside the bulk phases. In dimensional units it reads

$$\frac{\partial \mu^{\mathrm{out}}}{\partial t} = D_c \nabla^2 \mu^{\mathrm{out}} \; , \qquad (5.142)$$

where $D_c \equiv \Gamma_c (\partial \mu^{\mathrm{out}}/\partial c^{\mathrm{out}})_{\mathrm{eq}}$ is a diffusion constant. The value of D_c depends on the bulk equilibrium phase considered.

The "Gibbs Surface Solvability Condition" To solve (5.142), initial values $\delta\mu_1^{\mathrm{out}}(\tilde{u} = 0, \tilde{s})$ are required. These result from matching to the inner solution $\delta\mu_1^{\mathrm{in}}(\tilde{u}, \tilde{s})$ at $\tilde{u} = \tilde{\zeta}$. To obtain $\delta\mu_1^{\mathrm{out}}(\tilde{u} = 0^{\pm}, \tilde{s})$ from $\delta\mu_1^{\mathrm{out}}(\tilde{u} = \pm\tilde{\zeta}, \tilde{s})$ with $\tilde{\zeta} \equiv \epsilon\zeta$, it is useful to Taylor expand around $\tilde{u} = \tilde{\zeta}$

$$\delta\mu_1^{\mathrm{out}}(\tilde{u}, \tilde{s}) = \delta\mu_1^{\mathrm{out}}(\pm\tilde{\zeta}, \tilde{s}) + (\tilde{u} \mp \tilde{\zeta})\frac{\partial \delta\mu_1^{\mathrm{out}}}{\partial \tilde{u}}\Big|_{\pm\tilde{\zeta}} + \cdots \; . \qquad (5.143)$$

In the outer region, $\tilde{\zeta} \ll 1$. This expansion is valid at $\tilde{u} = 0$

$$\delta\mu_1^{\mathrm{out}}(\pm\tilde{\zeta}, \tilde{s}) = \delta\mu_1^{\mathrm{out}}(0, \tilde{s}) \pm \tilde{\zeta}\frac{\partial \delta\mu_1^{\mathrm{out}}}{\partial \tilde{u}}\Big|_{\pm\tilde{\zeta}} + \cdots \; . \qquad (5.144)$$

Since $\delta\mu_1^{\mathrm{out}}(\pm\tilde{\zeta}) = \delta\mu_1^{\mathrm{in}}(\pm\bar{\zeta})$, one can use (5.137) and (5.144) to obtain

$$\begin{aligned} \Delta c\delta\mu_1^{\mathrm{out}}(0, \tilde{s}) = &-\frac{\sigma_c \bar{\kappa}}{\xi} + A_1 - \frac{v_1(\tilde{s})\xi^2}{\tau\Gamma_c}\int_{-\infty}^{\infty} d\bar{u}[c_0^{\mathrm{in}}(\bar{u}) - c_0^{\mathrm{out}}(\bar{u})]^2 \\ &- \frac{\partial \delta\mu_1^{\mathrm{in}}}{\partial \bar{u}}\Big|_{-\bar{\zeta}}\int_{-\infty}^{0} d\bar{u}[c_0^{\mathrm{out}}(\bar{u}) - c_0^{\mathrm{in}}(\bar{u})] \\ &- \frac{\partial \delta\mu_1^{\mathrm{in}}}{\partial \bar{u}}\Big|_{\bar{\zeta}}\int_{0}^{\infty} d\bar{u}[c_0^{\mathrm{out}}(\bar{u}) - c_0^{\mathrm{in}}(\bar{u})] \\ &+ \Delta c\frac{v_1(\tilde{s})\xi^2}{\tau\Gamma_c}\int_{0}^{\pm\infty} d\bar{u}[c_0^{\mathrm{out}}(\bar{u}) - c_0^{\mathrm{in}}(\bar{u})] \; . \end{aligned} \qquad (5.145)$$

This yields the appropriate boundary value for $\delta\mu_1^{\mathrm{out}}(0, \tilde{s})$. The inner solution $\delta\mu_1^{\mathrm{in}}(0)$ differs from the outer solution $\delta\mu_1^{\mathrm{out}}(0)$, since the matching is done at $\tilde{u} = \tilde{\zeta}$ and extrapolated linearly to $\tilde{u} = 0$ by (5.144). This extrapolation of $\delta\mu_1^{\mathrm{in}}$ to $\delta\mu_1^{\mathrm{out}}$ is illustrated in Fig. 5.6. It is essential that $\delta\mu_1^{\mathrm{out}}$ and $\delta\mu_1^{\mathrm{in}}$ coincide at $\tilde{u} = \tilde{\zeta}$, which is in analogy to taking the *thin interface limit* rather than the *sharp interface limit* discussed in the context of the example of the previous section. The right-hand side of (5.145) appears to depend on whether the inner and outer solutions are matched at $u = \zeta$ or $u = -\zeta$. This ambiguity is eliminated by defining the interface to be a Gibbs surface at $u = 0$, given by the condition

$$\int_{-\infty}^{\infty} du[c_0^{\mathrm{out}}(u) - c_0^{\mathrm{in}}(u)] = 0 \; . \qquad (5.146)$$

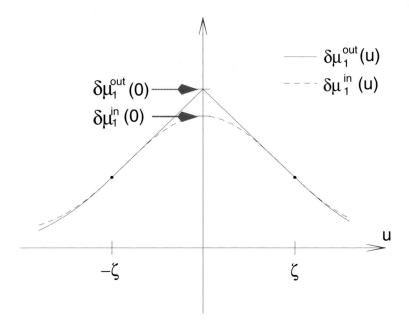

Fig. 5.6. Matching in the context of the generalized asymtotic analysis applied to $\delta\mu_1^{\text{in}}(u)$ (dashed line) with $\delta\mu_1^{\text{out}}(u)$ (solid line) at $u = \pm\zeta$.

It is always possible to satisfy (5.146) by choosing the position of the interface (i.e. $u = 0$) accordingly. In essence, the interface position is determined by the condition that the excess concentration on one side of the interface is exactly compensated by the deficit on the other, as depicted in Fig. 5.7. This can be interpreted as a solvability condition. It is one of the relations, which is required to obtain sharp interface model equations via this generalized asymptotic expansion.

A further necessary relation is obtained by matching the first derivative of the chemical potential. In dimensional units it reads:

$$v\Delta c = -\Gamma_c\left(\frac{\partial\delta\mu^{\text{in}}}{\partial u}\bigg|_{\zeta} - \frac{\partial\delta\mu^{\text{in}}}{\partial u}\bigg|_{-\zeta}\right). \tag{5.147}$$

Matching derivatives of the inner and outer solutions for μ and extrapolating to $u = 0^{\pm}$ by (5.143), yields the standard result

$$v\Delta c = -\Gamma_c\left(\frac{\partial\mu^{\text{out}}}{\partial u}\bigg|_{0+} - \frac{\partial\mu^{\text{out}}}{\partial u}\bigg|_{0-}\right), \tag{5.148}$$

since $\mu^{\text{out}}(u)$ is linear for $0 < |u| \leq \zeta$. Finally, combining (5.145), (5.146) and (5.148), results in the chemical potential of a moving, curved interface

$$\Delta c[\mu^{\text{out}}(0,s) - \mu_{\text{eq}}] = -\sigma_c\kappa + \mathcal{E}^2 v + A_1 + \mathcal{O}(\epsilon^2), \tag{5.149}$$

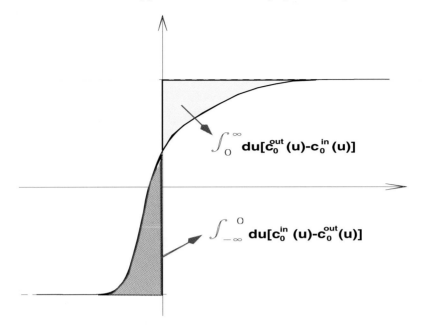

Fig. 5.7. Gibbs surface as defined by (5.146) pictured for $u = 0$. It matches the concentration deficit on one side with the concentration excess on the other.

where A_1 is given in (5.133) and

$$\mathcal{E}^2 \equiv \frac{1}{\Gamma_c} \int_{-\infty}^{\infty} du \left([c_0^{\text{out}}(u)]^2 - [c_0^{\text{in}}(u)]^2 \right) . \qquad (5.150)$$

All of the above can be combined into a single set of boundary layer equations written in terms of the concentration. In the outer region these boundary layer equations are related at order $\mathcal{O}(\varepsilon)$ to the chemical potential by the relationship

$$\delta\mu = \left. \frac{\partial\mu}{\partial c} \right|_{\text{eq}} \delta c . \qquad (5.151)$$

Combining this result with (5.149) and (5.129) yields the Gibbs–Thomson relation. In dimensionless units it reads

$$\frac{\delta c(0, s)}{\Delta c} = -d_o \kappa(s) - \beta_{\text{kin}} v , \qquad (5.152)$$

where d_o is the capillary length given by

$$d_o = \frac{\sigma}{(\Delta c)^2 (\partial\mu/\partial c)_{\text{eq}}} , \qquad (5.153)$$

$\sigma \equiv \sigma_c + \sigma_\phi$ the total surface tension given by

$$\sigma = \int_{-\infty}^{\infty} du \left[K_\phi \left(\frac{\partial \phi_0^{\text{in}}}{\partial u} \right)^2 + K_c \left(\frac{\partial c_0^{\text{in}}}{\partial u} \right)^2 \right] , \tag{5.154}$$

and β_{kin} is the coefficient of kinetic undercooling, which reads

$$\beta_{\text{kin}} = \frac{1}{(\Delta c)^2 (\partial \mu / \partial c)_{\text{eq}}} \left[\frac{\sigma_\phi}{K_\phi \Gamma_\phi} - \mathcal{E}^2 \right] . \tag{5.155}$$

Equation (5.152) provides a boundary condition at the interface for the diffusion equation (5.142), which can be written

$$\frac{\partial \delta c}{\partial t} = D_c \nabla^2 \delta c , \tag{5.156}$$

where

$$D_c \equiv \Gamma_c \left. \frac{\partial \mu}{\partial c} \right|_{\text{eq}} . \tag{5.157}$$

This must be solved in conjunction with (5.148), which can be transformed into

$$\Delta c \, v(s) = \left[D_c \frac{\partial \delta c}{\partial u} \right]_{0-} - \left[D_c \frac{\partial \delta c}{\partial u} \right]_{0+} . \tag{5.158}$$

The latter represents the well known Stefan condition, ensuring conservation of energy at the interface. Thus (5.152)–(5.158) recover the sharp interface formulation of this general class of moving interface problems represented by an energy functional $\mathcal{F}$ as given in (5.105). Equations (5.153)–(5.155) relate the physical parameters of the sharp interface description to the phenomenological parameters of the diffuse interface model equations. They are determined by K_ϕ, K_c and by the exact position of the interface $u = 0$, as expressed by $\mathcal{E}^2$. However, there remains a freedom to choose f in such a way that thermodynamic consistency can be guaranteed.

5.4 Discussion

Here I will briefly summarize the impact of asymptotic matching approaches for the overall context of diffuse interface modeling. Such methods can serve to establish the correspondence of diffuse interface models and sharp interface models. In this sense an asymptotic analysis does not necessarily involve a discussion of thermodynamic consistency. Instead the latter model is validated completely by its equivalence to the sharp interface model in the respective limits (either the "sharp" or the "thin interface limit"). However, such a procedure misses the chance to apply diffuse interface modeling to growth phenomena for which sharp interface formulations are not yet established. In this case the thermodynamic theory underlying diffuse interface models opens the possibility to proceed in the opposite direction, i.e. formulate the diffuse interface model on the basis of a variation of the thermodynamic potential(s) first and then derive the sharp interface model from the latter through the asymptotic analysis by at the same time taking into account additional conditions to ensure strict relaxation of the potentials.

In this sense the great advance of the generalized asymptotic analysis as described in Sect. 5.3 is to ensure thermodynamic consistency inherently through general algebraic operations, which are independent of the specific form of the free energy functional[5]. Thus the calculations are universal in the sense that a freedom to choose $\mathcal{F}$ remains. Exploiting the concept of the Gibbs surface explained in detail in Sect. 5.3.2 yields a solvability condition to carry out the matching within this generalized framework.

Consequently this generalized asymptotic analysis opens the way to construct physically consistent sharp interface descriptions of more complicated multiple phase systems such as a solid in contact with a fluid which can support flows. This will involve mode coupling terms in the dynamical equations. Once such governing equations are constructed, there is no longer any conceptual difficulty in deriving the corresponding sharp interface equations. In Chap. 6 I will present a detailed discussion of growth influenced by diffusive as well as hydrodynamic transport in the bulk phase following these ideas of exploiting the diffuse interface approach to establish the precise sharp interface boundary conditions of the system.

5.5 Appendix

Appendix A: Transformation to Curvilinear Coordinates

The curvilinear coordinates (u, s) used in the text are related to the Cartesian coordinates as [103]

$$\mathbf{r} = \mathbf{R}(s) + u\hat{\mathbf{n}}(s) \,, \tag{5.159}$$

where $\mathbf{R}$ is the position of the interface and $\hat{\mathbf{n}}(s)$ is the normal vector. The metric tensor $g_{\alpha\beta}$ of the transformation from Cartesian to curvilinear coordinates is given by

$$g = \begin{pmatrix} 1 & 0 \\ 0 & (1 + u\kappa)^2 \end{pmatrix} \,,$$

where $\kappa = \partial\theta/\partial s$ is the curvature with θ being the angle between the r^1-axis and the tangent to the curve. The Laplacian in (u, s) is obtained in the following manner

$$\nabla^2 = \sum_{\alpha,\beta} \frac{1}{\sqrt{|g|}} \frac{\partial}{\partial r^\alpha} \sqrt{|g|} g^{\alpha\beta} \frac{\partial}{\partial r^\beta}$$

$$= \frac{\partial^2}{\partial u^2} + \frac{\kappa}{1 + u\kappa} \frac{\partial}{\partial u} + \frac{1}{(1 + u\kappa)^2} \frac{\partial^2}{\partial s^2} - \frac{u\kappa_s}{(1 + u\kappa)^3} \frac{\partial}{\partial s} \,, \tag{5.160}$$

---

[5] This holds provided that $\mathcal{F}$ describes well defined phases.

where $r^1 = u$ and $r^2 = s$. (r^1 and r^2 denote the components of $\mathbf{r}$.) $g^{\alpha\beta}$ are the components of the inverse of the matrix g and $\kappa_s \equiv \partial\kappa/\partial s$.

Within the inner region, the fields vary more rapidly in u direction than in s direction. This allows us to rescale coordinates (u, s) to dimensionless variables as $(\bar{u}, \bar{s}) \equiv (u/\xi, \varepsilon s/\xi)$. The dimensionless curvature is expressed by $\bar{\kappa} = \xi\kappa/\varepsilon$ and $\bar{\kappa}_{\bar{s}} = \xi^2\kappa_s/\varepsilon^2$, respectively. In terms of these dimensionless variables the Laplacian reads

$$\xi^2 \nabla^2 = \frac{\partial^2}{\partial\bar{u}^2} + \frac{\varepsilon\bar{\kappa}}{1 + \epsilon\bar{u}\bar{\kappa}}\frac{\partial}{\partial\bar{u}} + \frac{\epsilon^2}{(1 + \epsilon\bar{u}\bar{\kappa})^2}\frac{\partial^2}{\partial\bar{s}^2} - \frac{\epsilon^3\bar{u}\bar{\kappa}_{\bar{s}}}{(1 + \epsilon\bar{u}\bar{\kappa})^3}\frac{\partial}{\partial\bar{s}}$$

$$= \frac{\partial^2}{\partial\bar{u}^2} + \bar{\kappa}\epsilon\sum_{n=0}(-\epsilon\bar{u}\bar{\kappa})^n\frac{\partial}{\partial\bar{u}} + \left(\epsilon^2\sum_{n=0}(n+1)(-\epsilon\bar{u}\bar{\kappa})^n\right)\frac{\partial^2}{\partial\bar{s}^2}$$

$$- \frac{\bar{u}\bar{\kappa}_{\bar{s}}}{2}\left(\epsilon^3\sum_{n=0}(n+1)(n+2)(-\epsilon\bar{u}\bar{\kappa})^n\right)\frac{\partial}{\partial\bar{s}}$$

$$= \mathcal{L}_0 + \epsilon\mathcal{L}_1 + \epsilon^2\mathcal{L}_2 + \epsilon^3\mathcal{L}_3 + \cdots , \tag{5.161}$$

where

$$\mathcal{L}_0 = \partial^2/\partial\bar{u}^2 \tag{5.162}$$
$$\mathcal{L}_1 = \bar{\kappa}\partial/\partial\bar{u} \tag{5.163}$$
$$\mathcal{L}_2 = \partial^2/\partial\bar{s}^2 - \bar{\kappa}^2\bar{u}\partial/\partial\bar{u} \tag{5.164}$$
$$\mathcal{L}_3 = -2\bar{u}\bar{\kappa}\partial^2/\partial\bar{s}^2 + \bar{\kappa}^3\bar{u}^2\partial/\partial\bar{u} - \bar{u}\bar{\kappa}_{\bar{s}}\partial/\partial\bar{s} . \tag{5.165}$$

Appendix B: Greens Functions

In addition an expansion for the inverse of the Laplacian, i.e. the Greens function, is required [103]. The Greens function of interest is defined by

$$\nabla_{\mathbf{r}}^2 G(\mathbf{r}, \mathbf{r}') = \delta(\mathbf{r} - \mathbf{r}') . \tag{5.166}$$

An expansion of this function can be derived in the following manner. Let $G(\mathbf{r}, \mathbf{r}') = G_0(\bar{u}, \bar{s}; \bar{u}', \bar{s}') + \varepsilon G_1(\bar{u}, \bar{s}; \bar{u}', \bar{s}') + \cdots$, where

$$\mathcal{L}_0 G_0 = 0 \tag{5.167}$$
$$\mathcal{L}_0 G_1 + \mathcal{L}_1 G_0 = \delta(\bar{u} - \bar{u}')\delta(\bar{s} - \bar{s}') \tag{5.168}$$
$$\mathcal{L}_0 G_2 + \mathcal{L}_1 G_1 + \mathcal{L}_2 G_0 = 0 \tag{5.169}$$

$$\cdots .$$

The solution for G_0 is 0. Thus to lowest order the relevant solution for G is G_1. G_1 satisfies the equation

$$\frac{\partial^2}{\partial\bar{u}^2}G_1(\bar{u}, \bar{s}; \bar{u}', \bar{s}') = \delta(\bar{u} - \bar{u}')\delta(\bar{s} - \bar{s}') , \tag{5.170}$$

with solutions

$$G_1^-(\bar{u}, \bar{s}; \bar{u}', \bar{s}') = \begin{cases} \bar{u}\delta(\bar{s} - \bar{s}') & -\zeta < \bar{u}' < \bar{u} < 0 \\ \bar{u}'\delta(\bar{s} - \bar{s}') & -\zeta < \bar{u} < \bar{u}' < 0 \,, \end{cases}$$

$$G_1^+(\bar{u}, \bar{s}; \bar{u}', \bar{s}') = \begin{cases} -\bar{u}'\delta(\bar{s} - \bar{s}') & 0 < \bar{u}' < \bar{u} < \zeta \\ -\bar{u}\delta(\bar{s} - \bar{s}') & 0 < \bar{u} < \bar{u}' < \zeta \,. \end{cases}$$

The surface terms $\mathcal{S}^\pm$ of (5.115) can be expanded as

$$\mathcal{S}^\pm = \pm \int_0^{\pm\bar{\zeta}} d\bar{u} \frac{\partial c_0^{in}}{\partial \bar{u}} \oint_B d\bar{s}' \left[\delta\mu(\bar{u}', \bar{s}') \frac{\partial G^\pm}{\partial \bar{u}'} - G^\pm \frac{\partial \delta\mu(\bar{u}', \bar{s}')}{\partial \bar{u}'} \right] \Big|_B$$

$$= \pm \int_0^{\pm\bar{\zeta}} d\bar{u} \frac{\partial c_0^{in}}{\partial \bar{u}} \oint_B d\bar{s}' \sum_{n=1} \sum_{m=0} \epsilon^{n+m-1} \times \left[\delta\mu_n(\bar{u}', \bar{s}') \frac{\partial G_m^\pm(\bar{u}, \bar{s}; \bar{u}', \bar{s}')}{\partial \bar{u}'} \right.$$

$$\left. - G_m^\pm(\bar{u}, \bar{s}; \bar{u}', \bar{s}') \frac{\partial \delta\mu_n(\bar{u}', \bar{s}')}{\partial \bar{u}'} \right] \Big|_B = \epsilon \mathcal{S}_1^\pm + \epsilon^2 \mathcal{S}_2^\pm + \cdots \,, \tag{5.171}$$

where

$$\mathcal{S}_1^\pm = \pm \int_0^{\pm\bar{\zeta}} d\bar{u} \frac{\partial c_0^{in}}{\partial \bar{u}} \oint_B d\bar{s}' \left(\delta\mu_1 \frac{\partial G_1^\pm}{\partial \bar{u}'} - G_1^\pm \frac{\partial \delta\mu_1}{\partial \bar{u}'} \right) \Big|_B$$

$$\mathcal{S}_2^\pm = \pm \int_0^{\pm\bar{\zeta}} d\bar{u} \frac{\partial c_0^{in}}{\partial \bar{u}} \oint_B d\bar{s}' \left[\delta\mu_2 \frac{\partial G_1^\pm}{\partial \bar{u}'} + \delta\mu_1 \frac{\partial G_2^\pm}{\partial \bar{u}'} \right.$$

$$\left. - G_2^\pm \frac{\partial \delta\mu_1}{\partial \bar{u}'} - G_1^\pm \frac{\partial \delta\mu_2}{\partial \bar{u}'} \right] \Big|_B \,. \tag{5.172}$$

Here the subscript B indicates that the integrands are evaluated on the boundary at $\bar{u}' = 0^\pm$ and at $\bar{u}' = \pm\bar{\zeta}$.

Moreover the $\mathcal{O}(\varepsilon)$ surface contribution becomes:

$$\mathcal{S}_1 = \mathcal{S}_1^- + \mathcal{S}_1^+$$

$$= \int_{-\bar{\zeta}}^0 d\bar{u} \frac{\partial c_0^{in}}{\partial \bar{u}} \left(\delta\mu_1^{in}(0, \bar{s}) + \bar{u} \frac{\partial \delta\mu_1^{in}(\bar{u}', \bar{s})}{\partial \bar{u}'} \Big|_{-\bar{\zeta}} \right)$$

$$+ \int_0^{+\bar{\zeta}} d\bar{u} \frac{\partial c_0^{in}}{\partial \bar{u}} \left(\bar{u} \frac{\partial \delta\mu_1^{in}(\bar{u}', \bar{s})}{\partial \bar{u}'} \Big|_{+\bar{\zeta}} + \delta\mu_1^{in}(o, \bar{s}) \right)$$

$$= \delta\mu_1^{in}(0, \bar{s}) + \frac{\partial \delta\mu_1^{in}}{\partial \bar{u}'} \Big|_{-\bar{\zeta}} \int_{-\bar{\zeta}}^0 d\bar{u} [c_0^{in}(-\bar{\zeta}) - c_0^{in}(\bar{u})]$$

$$+ \frac{\partial \delta\mu_1^{in}}{\partial \bar{u}'} \Big|_{+\bar{\zeta}} \int_0^{\bar{\zeta}} d\bar{u} [c_0^{in}(+\bar{\zeta}) - c_0^{in}(\bar{u})] \,. \tag{5.173}$$

To evaluate the bulk contribution $\mathcal{B}$, (5.114) is expanded in powers of ε. Then (5.114) reads

$$\mathcal{B}^\pm = \pm \frac{1}{\Gamma_c} \int_0^{\pm\zeta} du \frac{\partial c_0^{in}}{\partial u} \int_{V_\pm} d\mathbf{r}' G^\pm(\mathbf{r}, \mathbf{r}') \frac{\partial c^{in}(\mathbf{r}', t)}{\partial t} . \tag{5.174}$$

Further a variable transformation $\mathbf{r}' \to \bar{u}', \bar{s}'$ yields

$$\begin{aligned}
\mathcal{B}^\pm &= \pm \frac{\xi^2}{\tau\Gamma_c} \int_0^{\pm\bar\zeta} d\bar{u} \int_{V_\pm} \frac{d\bar{u}' d\bar{s}'}{1 + \bar{u}' \epsilon \bar\kappa(\bar{s}')} \sum_{n=1}^\infty \sum_{m=1}^\infty \epsilon^{n+m-1} \\
&\quad \times v_n G_m^\pm(\bar{u}, \bar{s}; \bar{u}', \bar{s}') \frac{\partial c_0^{in}}{\partial \bar{u}} \frac{\partial}{\partial \bar{u}'} (c_0^{in} + \epsilon \delta c_1^{in} + \cdots) \\
&= \epsilon \mathcal{B}_1^\pm + \epsilon^2 \mathcal{B}_2^\pm + \cdots ,
\end{aligned} \tag{5.175}$$

where

$$\begin{aligned}
\mathcal{B}_1^\pm &= \pm \frac{\xi^2}{\tau\Gamma_c} \int_0^{\pm\bar\zeta} d\bar{u} \int_{V_\pm} d\bar{u}' d\bar{s}' v_1 G_1^\pm \frac{\partial c_0^{in}}{\partial \bar{u}} \frac{\partial c_0^{in}}{\partial \bar{u}'} \\
\mathcal{B}_2^\pm &= \pm \frac{\xi^2}{\tau\Gamma_c} \int_0^{\pm\bar\zeta} d\bar{u} \int_{V_\pm} d\bar{u}' d\bar{s}' \Big[(v_2 G_1^\pm + v_1 G_2^\pm \\
&\quad - v_1 \bar\kappa \bar{u}' G_1^\pm) \frac{\partial c_0^{in}}{\partial \bar{u}} \frac{\partial c_0^{in}}{\partial \bar{u}'} + v_1 G_1^\pm \frac{\partial c_0^{in}}{\partial \bar{u}} \frac{\partial \delta c_1^{in}}{\partial \bar{u}'} \Big] .
\end{aligned} \tag{5.176}$$

For the Greens function introduced above, $\mathcal{B}_1$ results in

$$\mathcal{B}_1 = \mathcal{B}_1^- + \mathcal{B}_1^+ = -\frac{\xi^2 v_1(\bar{s})}{\tau\Gamma_c} \int_{-\bar\zeta}^{+\bar\zeta} d\bar{u} [c_0^{in}(\bar{u}) - c_0^{out}(\bar{u})]^2 . \tag{5.177}$$

6. Application of Diffuse Interface Modeling to Hydrodynamically Driven Growth

Taking up the line of argumentation of the previous chapter, here I will start with a diffuse interface model of hydrodynamically driven growth at solid–liquid interfaces. In particular I will be interested in the dendritic morphology which is known to occur for quenches taking the system under investigation from the liquid region of the underlying phase diagram to the liquid/solid coexistence region (see Fig. 5.3 and Fig. 5.4). Basically dendritic growth can be pictured as the advance of a solid phase shaped like a *dendrite* (see Fig. 6.1) into the liquid phase. One question tied to this growth phenomenon is that of the precise shape and the precise velocity selected during the growth process. This is referred to as the *selection problem of dendritic growth*. Here I will reconsider the selection problem of the dendritic morphology starting from a thermodynamically consistent phase-field model of hydrodynamically influenced dendritic growth. I will take its thermodynamic consistency as validation of the model. I will then continue to carry out the asymptotic analysis of the diffuse interface model. This analysis reveals an additional term beyond the ones classically known within the sharp interface equations describing dendritic growth. Due to the thermodynamic consistency of the diffuse interface model from which it is derived I claim this additional term to reveal further insight into the physics at the dendritic interface which has not been considered classically. This is relevant for the comparison of classical dendritic growth theory and related experiments, which still reveals unsettled issues. Within this chapter I will recall these issues and meet them with an analysis of the 'new' model equations including the non-classical term obtained only by the analysis of the related thermodynamically consistent diffuse interface model in the beginning.

Classically the problem is treated based on a set of equations containing the Navier–Stokes equations and the relevant diffusion equations in the bulk phases. These have to be supplemented by the correct boundary conditions given by the Gibbs–Thomson relation, the Stefan condition, the no-slip condition and the condition for the conservation of mass at the interface. The basic structure of this set of equations will remain unchanged within this chapter, however an additional curvature term within the Stefan condition implies an energetic contribution in the interfacial region relevant for further consideration.

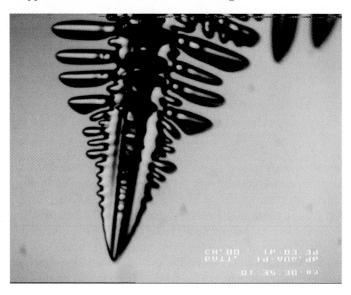

Fig. 6.1. Picture of a single dendrite. Its precise shape and velocity are the main issues of this chapter. This dendrite was grown at the Laboratory of Solid State Physics of the ETH Zürich in the group of Prof. Bilgram. Picture kindly provided by Stalder.

A couple of diffuse interface models for solid–liquid interfaces including hydrodynamic transport in the bulk can be found in the literature so far [12, 13, 15, 17, 85, 250]. The basic assumption underlying their formulation is that the solid phase is well approximated by a highly viscous isotropic liquid. This requires the inclusion of Korteweg stresses, which can drive convection [190]. Korteweg stresses are known from the theory of fluids near a critical point, where they are given by a capillary tensor acting as the reversible part of the stress tensor[1]. Having in mind the structure of diffuse interface models for diffusion limited growth, the basic formalism of models extended to hydrodynamic transport arises from the transformation

$$\frac{\delta}{\delta t} \rightarrow \frac{\mathrm{D}}{\mathrm{D} t} := \frac{\delta}{\delta t} + \mathbf{v} \cdot \nabla \, , \qquad (6.1)$$

where $\mathbf{v}$ is the vector of the fluid velocity field. Moreover, consistency with a formulation per unit mass rather than per unit volume has to be ensured, since obviously mass is a conserved quantity whereas local density is not. One can recognize the quantity on the right hand side of (6.1) as the material derivative with respect to time, i.e. the phase-field equations become invariant under Galilean transformation.

---

[1] The irreversible part is provided by the standard viscous stress term of a Newtonian liquid.

The diffuse interface model, which underlies the analysis of this chapter, was presented before at the end of Chap. 4. An asymptotic analysis of the model was given by Anderson *et al.* in [18]. Here I will recall their argumentation and summarize some of their results as starting point for the reconsideration of the dendritic selection problem with density change flow included.

6.1 Diffuse Interface Modeling for Hydrodynamically Influenced Growth

The formulation of the headline of this section reflects some of the difficulties one encounters when talking about growth of a phase-separating interface taking into account hydrodynamic transport in the liquid bulk phase. Whereas "diffusion limited" growth is a fixed expression, "convection limited" growth does not seem to describe the possible physical set-up belonging to that expression correctly: usually hydrodynamic as well as diffusive transport of the temperature field[2] will be active in the bulk. In most cases diffusive transport will even be the dominant one with hydrodynamic transport only imposing corrections. Nevertheless it remains challenging to fully grasp these corrections quantitatively for the big range of applications appearing in e.g. solidification, thin film growth, epitaxial growth or electrodeposition, to mention only a few. Modeling certainly is the first step to deal with these problems. Within the framework of diffuse interface modeling it was demonstrated at the end of Chap. 4 – following [14] – how a suitable model can be derived by expressing entropy production in terms of transport variables and identifying the forms which ensure entropy production to be non-negative. One important feature of this model is that it couples fluid motion to a non-conserved order parameter description. This offers the possibility to treat quasi-incompressible systems [208], in which the densities of the two bulk phases are each spatially uniform. However, the overall density ρ can be formulated as function of the order parameter varying smoothly within the interfacial region. Thus

$$\rho(\phi) = \rho_S r(\phi) + \rho_L [1 - r(\phi)] , \qquad (6.2)$$

where ρ_S and ρ_L are the densities in liquid and solid, respectively. The function $r(\phi)$ is monotonically increasing with $r(0) = 0$ and $r(1) = 1$. Suitable choices include $r(\phi) = \phi$ and $r(\phi) = \phi^2(3 - 2\phi)$. Assumption (6.2) is called quasi-incompressible, since density may vary in the interfacial region but thereby does not depend on pressure. It places a constraint on the form of the underlying thermodynamic potential as stated in [208]. In particular it requires that the underlying Gibbs free energy density is given by

---

[2] In case of multicomponent materials diffusion of solute will be involved in the growth process as well.

$$g(T, p, \phi) = g_0(T, \phi) + \frac{p - p_R}{\rho(\phi)} , \tag{6.3}$$

where p_R is a reference pressure associated with an isothermal, stationary planar interface at melting temperature T_M.

In their analysis in [18] Anderson *et al.* take $g_0(T, \phi)$ as

$$
\begin{aligned}
g_0(T, \phi) = {} & \left[e_0 - cT_M - r(\phi)L - \frac{1}{4a_S} H_m(\phi) \right] \left(1 - \frac{T}{T_M} \right) \\
& - c \cdot T \ln(\frac{T}{T_M}) + \frac{1}{4a} H_m(\phi) ,
\end{aligned}
\tag{6.4}
$$

in which case the corresponding expressions for the internal energy and entropy densities are

$$e = e_0 + c(T - T_M) - r(\phi)L + \frac{1}{4a_E} H_m(\phi) - \frac{p_R}{\rho(\phi)} \tag{6.5}$$

$$s = \frac{1}{T_M} [e_0 - r(\phi)L + \frac{1}{4a_S} H_m(\phi)] + c \cdot \ln(\frac{T}{T_M}) . \tag{6.6}$$

Here $1/a_E = 1/a - 1/a_S$. $1/a$ is simply the height of the double well of the Gibbs free energy density at $T = T_M$. $1/a_E$ and $1/a_S$ are heights of the double wells within the internal energy and entropy densities, respectively. Moreover, the quantity e_0 is a constant reference energy and both the heat capacity per unit mass c and the latent heat per unit mass L are assumed to be constant.

Note that the double–well function H_m appears in the "per unit mass" quantity $g_0(T, \phi)$. The analog per unit volume quantity associated with $\rho(\phi)g_0(T, \phi)$ is $\rho(\phi)H_m(\phi) \equiv H_v(\phi)$. A common form of a double–well potential is the function $\phi^2(1 - \phi)^2$. In standard phase-field models, which do not include convection, it is usually associated with a "per unit volume" quantity (see e.g. [289]). In applications, in which the density of the two phases is constant and equal, the per unit mass and the per unit volume specification of the double well are equivalent. However, in the present situation the bulk densities are not necessarily equal. Thus the per unit mass formula $H_m(\phi) = \phi^2(1 - \phi)^2$ and the per unit volume formula $H_v(\phi) = \rho_L \phi^2(1 - \phi)^2$ with subsequent $H_m(\phi) = \rho_L \phi^2(1 - \phi)^2/\rho(\phi)$ result in different approaches.

For a detailed description of the algebra involved in the asymptotic analysis of (4.98)–(4.101) the interested reader is referred to [18]. Here I will merely state the basic steps and results: Assuming an expansion of outer variables as

$$\mathbf{u} = \mathbf{u}_0 + \varepsilon \mathbf{u}_1 + \varepsilon^2 \mathbf{u}_2 + ... \tag{6.7}$$

$$p = p_0 + \varepsilon p_1 + \varepsilon^2 p_2 + ... \tag{6.8}$$

$$\phi = \phi_0 + \varepsilon \phi_1 + \varepsilon^2 \phi_2 + ... \tag{6.9}$$

$$\theta = \theta_0 + \varepsilon \theta_1 + \varepsilon^2 \theta_2 + ... \tag{6.10}$$

and an analogous expansion of inner variables as

$$\mathbf{U} = \mathbf{U}_0 + \varepsilon \mathbf{U}_1 + \varepsilon^2 \mathbf{U}_2 + \dots \tag{6.11}$$

$$P = \frac{1}{\varepsilon}[P_0 + \varepsilon P_1 + \varepsilon^2 P_2 + \dots] \tag{6.12}$$

$$\Phi = \Phi_0 + \varepsilon \Phi_1 + \varepsilon^2 \Phi_2 + \dots \tag{6.13}$$

$$\Theta = \Theta_0 + \varepsilon \Theta_1 + \varepsilon^2 \Theta_2 + \dots \tag{6.14}$$

$$\mathbf{M} = \frac{1}{\varepsilon}[\mathbf{M}_0 + \varepsilon \mathbf{M}_1 + \varepsilon^2 \mathbf{M}_2 + \dots] \tag{6.15}$$

leading-order equations in the bulk phases read:

$$\frac{D\mathbf{u}_0}{Dt} = -\nabla p_0 + Pr \nabla \cdot \tau_0 \tag{6.16}$$

$$\nabla \cdot \mathbf{u_0} = 0 \tag{6.17}$$

$$\frac{D\theta_0}{Dt} = k\nabla^2 \theta_0 + \frac{SPr}{\Lambda\tilde{\gamma}}\tau_0 \tag{6.18}$$

$$\phi_0 = \begin{cases} 0 & \text{in the liquid phase} \\ 1 & \text{in the solid phase .} \end{cases} \tag{6.19}$$

Here θ is related to the internal energy e via

$$\theta = e - r(\phi) + \frac{\delta}{2}H_m(\phi) - \frac{\varepsilon Sp^\star}{\tilde{\gamma}\lambda}\frac{1}{\rho} . \tag{6.20}$$

Apart from the last term on the right hand side of (6.18) the above set of equations (6.16)–(6.18) are nothing but the incompressible Navier–Stokes equations and the diffusion equation for the temperature field respectively. τ_0 denotes the leading-order terms of the viscous stress tensor given at the end of Chap. 4. Pr refers to the well known Prandtl number representing the ratio of kinematic viscosity to diffusivity in the material. The additional term $\frac{SPr}{\Lambda\tilde{\gamma}}\tau_0$ can be understood as a source term resulting from the rate of work of forces acting at the boundary of each volume element (compare to (4.80)). For small coefficients $\frac{Pr}{\tilde{\gamma}}$ this term can be neglected.

To obtain information about the physics in the interfacial region resulting asymptotically in boundary conditions, I turn to the analysis of the behavior of inner variables invoking matching and solvability conditions as described in the previous chapter:

Continuity Equation
The leading-order problem for the continuity equation (4.98) can be solved to obtain the form of U_0^3 through the interface as

$$U_0^3 = \frac{J_0}{\rho(\Phi_0)} + V_n , \tag{6.21}$$

where $J_0 = \rho_L(\mathbf{u}_0|_L \cdot \hat{\mathbf{n}} - V_n) = \rho_S(\mathbf{u}_0|_S \cdot \hat{\mathbf{n}} - V_n)$. Thus matching the velocities results in

$$\rho(\mathbf{u}_0 \cdot \hat{\mathbf{n}} - V_n)|_S^L = 0 \,, \tag{6.22}$$

which is nothing but a boundary condition stating conservation of mass through the interface.

Momentum Equation

As a second condition for the velocity components one would expect the so called no-slip condition for the tangential velocity components to arise at the interface. Indeed it can be recovered by an analysis of the momentum equation (4.99). At leading-order $[\mathcal{O}(\varepsilon^{-1})]$ one finds that the normal[3] derivatives of the velocity components satisfy

$$[\mathbf{u}_0 - \hat{\mathbf{n}}(\mathbf{u}_0 \cdot \hat{\mathbf{n}})]|_S^L = 0 \,. \tag{6.23}$$

Together with (6.22) this implies that the tangential components of the velocity field vanish.

Phase-Field Equation

Analyzing the phase-field equation (4.100) at leading- and at first-order yields a solvability condition, which can be expressed as

$$L(1 - \frac{T_I}{T_M}) = -\frac{\gamma_0}{\rho_L}\mathcal{K} - (\mathbf{u}_0 \cdot \hat{\mathbf{n}} - V_n)\frac{L}{T_M}\frac{1}{\mu_{\text{mob}}} \,. \tag{6.24}$$

Here T_I is the interface temperature and μ_{mob} the interface mobility. $\mathcal{K}$ represents the curvature of the interface in a two-dimensional system, γ_0 the surface energy. Thus (6.24) is nothing but the classical Gibbs–Thomson condition with the two terms on the right-hand side of the equation summarizing the effects of curvature and attachment kinetics, respectively.

Energy Equation

Last the energy equation (4.101) has to be taken into account. Leading order contributions originate from dissipation, from thermal diffusion and from the advective term in the energy equation. Examining the $\mathcal{O}(1)$ problem one can identify a heat flux boundary condition. The respective analysis involves the inner velocity component correction U_1^3, which is determined from the correction to the continuity equation. It yields a heat flux boundary condition as follows

$$k\hat{\mathbf{n}} \cdot \nabla T|_S^L = (\mathbf{u}_0|_L \cdot \hat{\mathbf{n}} - V_n)[\rho_L L + \gamma_0 \mathcal{K}] \,. \tag{6.25}$$

This is the classical Stefan condition modified to account for motion in the solid and liquid bulk phases as well as the curvature of the interface. The effect of flow only enters in the first factor on the right-hand side,

---

[3] normal with respect to the interface

where it arises as the normal velocity of the material relative to the interface. The curvature effect follows from the term $\xi_E^2 \nabla \cdot (\Gamma \Sigma) \frac{\mathrm{D}\Phi}{\mathrm{D}t}$ in (4.101). It represents the internal energy gradient and double–well terms in the model.

In [18] Anderson *et al.* discuss some earlier work considering this term: In the context of phase-field modeling it had been identified before by Fife and Penrose [116] and Fried and Gurtin [122]. Within sharp interface formulations Wollkind and Maurer [299], Umantsev and Davis [284], Zhang and Garimella [301] and Schlitz and Garimella [247] had studied its effect on the stability of an interface in the context of diffusion limited growth and Lemieux and Kotliar [205] its influence on velocity selection during diffusion limited dendritic growth. They showed that in this case the curvature term increases the values of selected velocities. Viewed in the light of experiments by Glicksman *et al.* summarized in 1994 in [129], the findings of Lemieux *et al.* dating back to 1987 appear quite interesting. Glicksman *et al.* researched the difference between the theory of diffusion and hydrodynamically influenced dendritic growth in detail based on microgravity experiments. Prior to these experiments Huang and Glicksman had performed terrestrial growth experiments using the transparent organic material succinonitrile (SCN) [149]. The data of the latter was in good agreement with the analytical theory for velocity selection in diffusion limited dendritic growth[4] except for low undercoolings. A plausible reason appeared to be the fact that at low undercoolings the driving force due to the temperature gradient gets smaller and smaller. In turn hydrodynamic effects start to become more and more important and could become visible exactly in such a manner that it would explain the deviations from the analytical curve. Microgravity experiments were thought to be the way out, since buoyancy – as main force driving hydrodynamic flow – is suppressed. Nevertheless the experiments by Glicksman *et al.* revealed deviations from the analytical curve at low undercoolings, as well, even though somewhat smaller in their absolute value (Fig. 6.2). Indeed there exists some flow, which occurs under microgravity as well, namely the so called *density change flow*. This expression density change flow refers to hydrodynamics with the only mechanism driving hydrodynamic flow being the density change the material undergoes when solidifying. Thus at this point one could wonder whether density change flow does lead to noticeable deviations from diffusion theory in particular at low undercoolings. However, steady state solutions of the classical sharp interface equations governing density change flow had been determined by McFadden and Coriell in 1986 [212]. They showed that the effect of density change flow is of the order of the density difference between solid and liquid phase normalized with respect to the density in the liquid, which I will call ϵ in the following:

---

[4] A thorough review discussing velocity selection in the case of diffusion limited dendritic growth is given in [42].

$$\epsilon = \frac{\rho_s - \rho_l}{\rho_l} \; .$$

This implies that it is minor for most experimental set-ups, in particular for SCN with $\epsilon = 0.028$. (Other values are, e.g., $\epsilon = 0.065$ for aluminium and $\epsilon = -0.09$ for silicon. Note that for silicon density increases upon solidification.) Moreover, in comparison to diffusion limited dendritic growth their analysis reveals a decrease of growth rates with ϵ. The experimental curve by Glicksman *et al.*, on the other hand, displays an increase (see Fig. 6.2). Thus even qualitatively density change flow proved to have rather the opposite effect than the one desired to explain the experiments. McFadden and Coriell did not proceed their studies with a perturbative analysis as necessary to investigate the influence of surface tension on the steady state solutions. A comprehensive solution including the dependence of growth rates on anisotropic surface tension β obtained via an irregular perturbation expansion was first given for diffusion limited dendritic growth by Brener [43]. It reveals corrections as $\sim \beta^{7/4}$. From this one can assume that for SCN with anisotropy factor $\beta = 0.55$ [224] considering surface tension effects within the McFadden/Coriell-solution cannot help to explain the experiments by Glicksman *et al.*. Thus at this point it seems useless to carry out the full irregular perturbation analysis for the classical problem of density change influenced dendritic growth hoping to gain new insight compatible with experimental evidence. However, turning from the classical sharp interface formulation of the problem towards the non-classical one with an additional curvature term in the energy equation as in (6.25), one can hope that the findings of Lemieux and Kotliar for diffusion limited dendritic growth will carry over, i.e. that also in the case of density change influenced dendritic growth an irregular perturbation expansion taking into account the additional curvature term will result in an increase of selected velocities and thereby explain the experiments.

The following section is devoted to an investigation of this idea. A first step is an analysis of (6.16)–(6.18) and boundary conditions (6.22)–(6.25) in the framework of dendritic selection theory. This analysis is carried out in a way originally proposed by Kruskal and Segur in [191]. In the context of dendritic growth this method was employed by various authors, see e.g. [236] and references therein. It starts from the derivation of a zero surface tension solution. This solution is subsequently analyzed further within the framework of regular as well as irregular perturbation theory.

6.2 The Selection Problem of Dendritic Growth Revisited

To begin a detailed study of dendritic selection in the case of density change flow at this point I recast the governing equations and boundary conditions suitably for the parameter regime $\frac{Pr}{\bar{\gamma}} \ll 1$ [117]. Here the operator $\frac{D}{Dt}$ is given

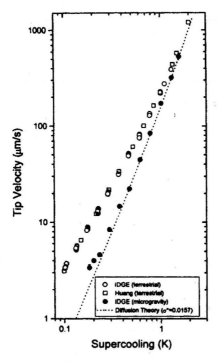

Fig. 6.2. Figure taken from [129]. The dotted line refers to the selection theory of diffusion limited dendritic growth, the solid data points to microgravity experiments and the white squares and circles to two different terrestrial experiments. The theory curve for density change influenced growth would run below the dotted curve depicting the diffusion theory (compare to Fig. 6.4).

by the full form $\frac{\delta}{\delta t}+\mathbf{u}\nabla$ to elucidate the necessary coordinate transformations in the following. Moreover, the thermal diffusivity D, the volumetric specific heat $c = \rho_s c_s = \rho_l c_l$ and $T_s c_s = T_l c_l$ are assumed equal in both phases.

Basic Governing Equations

In the **liquid phase** the heat conduction equation for the temperature field T corresponding to (6.18) can be written

$$\frac{\partial T}{\partial t} + (\mathbf{u}\cdot\nabla)\, T = D\,\nabla^2 T \, , \tag{6.26}$$

where $\mathbf{u}$ denotes the velocity of the hydrodynamic field satisfying mass conservation

$$\nabla \cdot \mathbf{u} = 0 \tag{6.27}$$

and the momentum equation

$$\frac{\partial \mathbf{u}}{\partial t} + (\mathbf{u}\cdot\nabla)\,\mathbf{u} = -\frac{1}{\rho_l}\nabla p + \nu\nabla^2\mathbf{u} \, . \tag{6.28}$$

ν is the kinematic viscosity of the material in the molten (liquid) state, ρ_l the density of the melt and p the pressure field. (6.26)–(6.28) hold for a frame of reference at rest as well as moving at constant velocity.

Within the **solid phase**, the heat conduction equation is of the same type as (6.26), however the velocity is constant and determined by the velocity of the reference frame. Choosing it as $\mathbf{V}_{\mathrm{fr}} = V\mathbf{e_x}$, one obtains

$$\frac{\partial T}{\partial t} + (V\mathbf{e_x} \cdot \nabla)\, T = D\, \nabla^2 T \,. \tag{6.29}$$

At the crystal–melt **interface** the fluid flow satisfies conservation of mass as well as a no-slip boundary condition:

$$\rho_l(\mathbf{u} - V\mathbf{e_x}) \cdot \mathbf{n} = (\rho_l - \rho_s)(\mathbf{v} - V\mathbf{e_x}) \cdot \mathbf{n}\,, \tag{6.30}$$

$$(\mathbf{u} - V\mathbf{e_x}) \cdot \mathbf{t} \;\; = 0\,, \tag{6.31}$$

where $\mathbf{v}$ denotes the velocity of the interface. These equations correspond to (6.22) and (6.23), respectively. $\mathbf{n}$ and $\mathbf{t}$ denote the unit normal and unit tangential vector to the moving interface. In (6.30) and (6.31) the velocity $\mathbf{u} - V\mathbf{e_x}$ refers to a frame of reference at rest. The temperature directly at the interface is specified claiming local thermodynamic equilibrium, which yields continuity of temperature

$$T_l = T_s \,, \tag{6.32}$$

as well as the Gibbs–Thomson relation

$$T = T_M \left(1 - \frac{\gamma}{\rho_s Q} \kappa \right) \,. \tag{6.33}$$

Equation (6.33) corresponds to (6.24) neglecting kinetic terms. Here, T_M is the crystallization temperature of the planar interface, $\rho_s Q$ the latent heat per unit volume of solid. It is related to the mean latent heat L appearing in (6.24) via $L = \frac{\rho_s}{\rho_l} \cdot Q$. γ denotes the anisotropic surface tension (see (2.8)) and κ the local curvature of the interface.

Moreover, energy is conserved for the phase transition occurring at the interface. Thus:

$$\rho_s Q \left(1 - \frac{\gamma \kappa}{\rho_s Q} \right) (\mathbf{u} - V\mathbf{e_x}) \cdot \mathbf{n} = [(k\nabla T)_s - (k\nabla T)_l] \cdot \mathbf{n} \,. \tag{6.34}$$

This is the final boundary condition at the interface corresponding to (6.25). The left-hand side of formula (6.34) expresses the latent heat release per unit volume of the solid phase. It is equal to the total energy flux away from the interface into both phases as given on the right-hand side of (6.34). Introducing a modified latent heat $\tilde{L}$, which takes into account unequal specific heats c_l and c_s as follows

$$\tilde{L} = Q + c_s T_M - c_l T_M \,, \tag{6.35}$$

it is possible to replace (6.34) by

$$\frac{\rho_s \tilde{L}}{\rho_l c_l}(1 - \frac{\gamma \kappa}{\rho_s Q})(\mathbf{u} - V\mathbf{e_x}) \cdot \mathbf{n} = D\left[(\nabla T)_s - (\nabla T)_l\right] \cdot \mathbf{n} . \qquad (6.36)$$

Boundary conditions at **infinity** $(x \to -\infty)$ are assumed to be given by:

$$\mathbf{u} \to V\mathbf{e_x} \qquad (6.37)$$

$$T \to T_\infty . \qquad (6.38)$$

Dimensionless Governing Equations

The analysis of the above set of equations (6.26)–(6.38) with respect to the selected growth velocities is simplified choosing dimensionless variables by rescaling spatial coordinates by ρ and the time coordinate by $\frac{\rho^2}{D}$.[5] Here ρ is a reference length scale specified as tip radius of curvature of the zero surface tension solution. The velocity field $\mathbf{u}$ is expressed in units of $\frac{D}{\rho}$, the temperature T in units of $\frac{\tilde{L}}{c_l}$ and pressure by $\frac{D^2 \rho_l}{\rho^2}$. Temperature is measured relative to the melting temperature of the flat interface T_M, i.e. $T = (T' - T'_M)/\frac{\tilde{L}}{c_l}$, where the prime indicates the dimensional temperature.

Expressed through dimensionless variables heat conduction in the **liquid phase** reads

$$\frac{\partial T}{\partial t} + (\mathbf{u} \cdot \nabla) T = \nabla^2 T . \qquad (6.39)$$

Mass conservation (6.27) remains formally unchanged. The momentum equation is now given by

$$\frac{\partial \mathbf{u}}{\partial t} + (\mathbf{u} \cdot \nabla) \mathbf{u} = -\nabla p + Pr \nabla^2 \mathbf{u} , \qquad (6.40)$$

where $Pr = \frac{\nu}{D}$ is the Prandtl number. To eliminate pressure one determines the curl of equation (6.40). This results in the vorticity equation

$$\frac{\partial \omega}{\partial t} + (\mathbf{u} \cdot \nabla) \omega = Pr \nabla^2 \omega . \qquad (6.41)$$

where $\omega = (\nabla \times \mathbf{u}) \cdot \mathbf{e_z}$ is the scalar vorticity. Heat conduction in the **solid phase** takes the dimensionless form:

$$\frac{\partial T}{\partial t} + P_c (\mathbf{e_x} \cdot \nabla) T = \nabla^2 T , \qquad (6.42)$$

---

[5] Actually the analysis of the phase-field model as summarized in the previous section is simplified by choosing dimensionless variables, as well. Moreover curvilinear coordinates were employed within the original work. To keep the summary of the previous section short such details were omitted. Again the interested reader is referred to [18].

where $P_c = \frac{\rho V}{D}$ is the Peclet number[6]. With $\frac{\rho_s}{\rho_l} = 1 + \epsilon$, the boundary conditions for the hydrodynamic field at the **interface** expressed in dimensionless variables read:

$$\mathbf{u} \cdot \mathbf{n} = P_c(1 + \epsilon)\mathbf{e_x} \cdot \mathbf{n} - \epsilon \mathbf{v} \cdot \mathbf{n} , \qquad (6.43)$$

$$\mathbf{u} \cdot \mathbf{t} = P_c \mathbf{e_x} \cdot \mathbf{t} . \qquad (6.44)$$

Condition (6.32) stating continuity of temperature remains unchanged. The Gibbs–Thomson equation now reads

$$T = -\delta a(\theta)\kappa , \qquad (6.45)$$

where $\delta = \frac{d_o}{\rho} = \frac{T'_M \gamma_o c}{Q^2 \rho}$ is the capillary length d_o in units of ρ.
    Finally, energy conservation can be expressed as

$$(1 + \epsilon - \delta\kappa)\left(-P_c\mathbf{e_x} + \mathbf{v}\right) \cdot \mathbf{n} = \left[(\nabla T)_s - (\nabla T)_l\right] \cdot \mathbf{n} . \qquad (6.46)$$

The boundary conditions at **infinity** become (6.38) and

$$\mathbf{u} \to P_c \mathbf{e_x} . \qquad (6.47)$$

Equations in Parabolic Coordinates

A further step to simplify the following analysis can be taken by transforming the above equations to parabolic coordinates (ξ, η) (see Fig. 6.3) defined by $x = \frac{1}{2}(\xi^2 - \eta^2)$ and $y = \xi\eta$ [219]. This transformation allows for a separation of variables. Moreover a stream function $\psi(\xi, \eta)$ defined as

$$w_\xi = \frac{1}{\sqrt{\eta^2 + \xi^2}} \frac{\partial \psi}{\partial \eta} \qquad (6.48)$$

$$w_\eta = \frac{-1}{\sqrt{\eta^2 + \xi^2}} \frac{\partial \psi}{\partial \xi} \qquad (6.49)$$

allows me to rewrite the governing bulk equations as follows:

Liquid Phase

$$(\xi^2 + \eta^2)\frac{\partial T}{\partial t} + \frac{\partial \psi}{\partial \eta}\frac{\partial T}{\partial \xi} - \frac{\partial \psi}{\partial \xi}\frac{\partial T}{\partial \eta} = \frac{\partial^2 T}{\partial \xi^2} + \frac{\partial^2 T}{\partial \eta^2} \qquad (6.50)$$

---

[6] Note that P_c is not exactly the same as the interfacial Peclet number $Pe_{\text{Int}} = \xi v/D$ (see Sect. 5.3) in the sense that the length ξ is related to the diffuse interface thickness of a phase-field model, whereas the length scale ρ denotes a characteristic length scale of the sharp interface model (6.26)–(6.38). Here it can be identified with the tip radius of the zero surface tension solution of the problem.

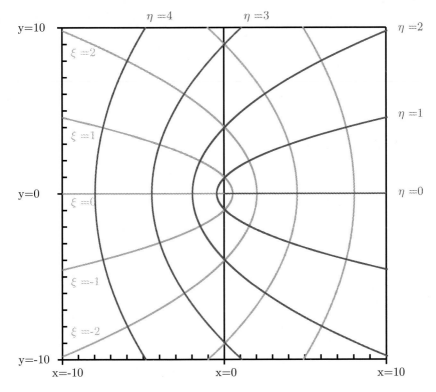

Fig. 6.3. Definition of parabolic coordinates (ξ, η) as underlying the analysis in the remainder of this section [219].

$$(\xi^2 + \eta^2)\frac{\partial \omega}{\partial t} + \frac{\partial \psi}{\partial \eta}\frac{\partial \omega}{\partial \xi} - \frac{\partial \psi}{\partial \xi}\frac{\partial \omega}{\partial \eta} = Pr\left(\frac{\partial^2 T}{\partial \xi^2} + \frac{\partial^2 T}{\partial \eta^2}\right) \qquad (6.51)$$

$$\frac{\partial^2 \psi}{\partial \xi^2} + \frac{\partial^2 \psi}{\partial \eta^2} + (\xi^2 + \eta^2)\omega = 0 \qquad (6.52)$$

Solid Phase

$$(\xi^2 + \eta^2)\frac{\partial T}{\partial t} + P_c\left(\xi\frac{\partial T}{\partial \xi} - \eta\frac{\partial T}{\partial \eta}\right) = \frac{\partial^2 T}{\partial \xi^2} + \frac{\partial^2 T}{\partial \eta^2} . \qquad (6.53)$$

Interface With the interface shape function η_s defined by $\eta - \eta_s(\xi, t) = 0$ one can rewrite the boundary conditions expressed in parabolic variables as follows:

• Mass Conservation

$$\partial_\xi \psi + \eta_s' \partial_\eta \psi = P_c(1 + \epsilon)\,(\xi\eta_s)' + \epsilon\left(\xi^2 + \eta_s^2\right)\partial_t\eta_s . \qquad (6.54)$$

• Continuity of tangential components of the hydrodynamic field

$$-\eta_s'\partial_\xi\psi + \partial_\eta\psi = P_c\left(\xi - \eta_s\eta_s'\right),\tag{6.55}$$

where the prime denotes differentiation with respect to ξ.

- Energy conservation

$$(1 + \epsilon - \delta\kappa)\left[P_c(\xi\eta_s)' + \left(\eta_s^2 + \xi^2\right)\partial_t\eta_s\right] = \left(-\eta_s'\partial_\xi + \partial_\eta\right)\left(T_s - T_l\right).\tag{6.56}$$

- Continuity of temperature (6.32)
- Gibbs–Thomson Relation

$$T = -\delta a[\eta_s(\xi)]\kappa[\eta_s(\xi)],\tag{6.57}$$

where the curvature is determined by

$$\kappa[\eta_s(\xi)] = -\frac{1}{\sqrt{\xi^2 + \eta_s^2}}\left(\frac{\eta_s''}{\left(1 + \eta_s'^2\right)^{\frac{3}{2}}} + \frac{\xi\eta_s' - \eta_s}{\left(\xi^2 + \eta_s^2\right)\left(1 + \eta_s'^2\right)^{\frac{1}{2}}}\right)\tag{6.58}$$

and anisotropy by

$$a[\eta_s(\xi)] = 1 - \beta + 8\beta\frac{\left(\xi\eta_s' + \eta_s\right)^2\left(\xi - \eta_s\eta_s'\right)^2}{\left(\xi^2 + \eta_s^2\right)^2\left(1 + \eta_s'^2\right)^2}.\tag{6.59}$$

Infinity Boundary conditions at infinity are now given by (6.38) and

$$\partial_\xi\psi \to P_c\eta,\tag{6.60}$$

$$\partial_\eta\psi \to P_c\xi.\tag{6.61}$$

(The underlying derivation of operators in parabolic coordinates can be found in the appendix of this chapter.)

Steady State Solutions

A first step when searching for the solutions to the problem given by (6.50) to (6.61) is to derive steady state solutions for the case of zero surface tension. These solutions can then be used as basis to find the solutions of the full problem in a perturbative manner. In the case of zero surface tension the temperature at the interface is constant ($T = 0$) and a family of exact solutions can be found. In order to determine these solutions one assumes that T depends on η only, so that the shape of the interface is a parabola $\eta_s(\xi) = 1$. In this case equation (6.50) reduces to:

$$-\frac{\partial\psi}{\partial\xi}\frac{\partial T}{\partial\eta} = \frac{\partial^2 T}{\partial\eta^2}.\tag{6.62}$$

Thus for the stream function the following ansatz

$$\psi = \xi f(\eta) \tag{6.63}$$

must hold. With (6.52) this yields a vorticity

$$\omega = -\frac{\xi}{\sqrt{\xi^2 + \eta^2}} f'' . \tag{6.64}$$

The simplest solution of (6.51) is the one for which the vorticity vanishes everywhere, i.e. $f'' = 0$. Integrating twice one finds that all boundary conditions can be met:

$$f(\eta) = P_c(\eta + \epsilon) . \tag{6.65}$$

Inserting relation (6.65) into (6.62) and integrating once results in

$$\frac{\partial T}{\partial \eta} = -P_c(1 + \epsilon) \exp\left[\frac{P_c}{2}(1 + \epsilon)^2\right] \exp\left[-\frac{P_c}{2}(\eta + \epsilon)^2\right] . \tag{6.66}$$

One further integration yields

$$T(\eta) = -P_c(1 + \epsilon) \exp\left[\frac{P_c}{2}(1 + \epsilon)^2\right] \int_1^\eta dx \, \exp\left[-\frac{P_c}{2}(x + \epsilon)^2\right] . \tag{6.67}$$

This can be expressed in terms of the complementary error function as

$$T(\eta) - T_\infty = \sqrt{\frac{\pi P_c}{2}}(1 + \epsilon) \exp\left[\frac{P_c}{2}(1 + \epsilon)^2\right] \text{erfc}\left(\sqrt{\frac{P_c}{2}}(\eta + \epsilon)\right) . \tag{6.68}$$

Moreover, the dimensionless undercooling can be obtained with $\eta = 1$ in (6.68) as:

$$\Delta = -T_\infty = \sqrt{\frac{\pi P_c}{2}}(1 + \epsilon) \exp\left[\frac{P_c}{2}(1 + \epsilon)^2\right] \text{erfc}\left(\sqrt{\frac{P_c}{2}}(1 + \epsilon)\right) . \tag{6.69}$$

As stated at the end of Sect. 6.1 steady state solutions of dendritic growth considering density change flow have been determined before by McFadden and Coriell in [212]. Their analysis was carried out in cylindric coordinates with same scaling, however defining P_c as $\frac{\rho V}{2D}$ rather than $\frac{\rho V}{D}$ as assumed here. The conclusion they draw from their analysis coincidences with (6.68), i.e. the effect of density change flow is of order ϵ and hence minor for most experimental set-ups. The remainder of this chapter is devoted to the question whether surface tension effects alter this preliminary view on dendritic growth influenced by density change flow noticeably. Within the context of classical sharp interface equations this analysis would have been useless as pointed out at the end of Sect. 6.1. However in connection with the results by Lemieux and Kotliar one can hope that the additional curvature term within the energy condition (6.34) will increase values of selected velocities in a way, that they provide an explanation for the experimental data in [129] at low undercoolings.

Regular Perturbation Expansion

A regular perturbation expansion serves to determine the stability of the above solution. To carry it out, the following expansions are employed:

$$T = T^0 + \sigma P_c(1+\epsilon)T^1 + \ldots \tag{6.70}$$

$$\eta_s = \eta_s^0 + \sigma\eta_s^1 + \ldots \tag{6.71}$$

$$\omega = \omega^0 + \sigma P_c\omega^1 + \ldots \tag{6.72}$$

$$\psi = \psi^0 + \sigma P_c\psi^1 + \ldots \;, \tag{6.73}$$

where $\sigma = \frac{\delta}{P_c(1+\epsilon)}$. Inserting these expansions the following set of first-order equations is obtained:

Liquid Phase

$$\frac{\partial^2 T^1}{\partial\xi^2} + \frac{\partial^2 T^1}{\partial\eta^2} - \frac{\partial\psi^0}{\partial\eta}\frac{\partial T^1}{\partial\xi} + \frac{\partial\psi^0}{\partial\xi}\frac{\partial T^1}{\partial\eta} = -\frac{\partial\psi^1}{\partial\xi}\frac{\partial T^0}{\partial\eta} \tag{6.74}$$

$$\frac{\partial^2 T^1}{\partial\xi^2} + \frac{\partial^2 T^1}{\partial\eta^2} - P_c\xi\frac{\partial T^1}{\partial\xi} + P_c(\eta+\epsilon)\frac{\partial T^1}{\partial\eta} = -\frac{\partial\psi^1}{\partial\xi}\frac{\partial T^0}{\partial\eta} \tag{6.75}$$

$$Pr\frac{\partial^2\omega^1}{\partial\xi^2} + Pr\frac{\partial^2\omega^1}{\partial\eta^2} - \frac{\partial\psi^0}{\partial\eta}\frac{\partial\omega^1}{\partial\xi} + \frac{\partial\psi^0}{\partial\xi}\frac{\partial\omega^1}{\partial\eta} = 0 \tag{6.76}$$

$$\frac{\partial^2\omega^1}{\partial\xi^2} + \frac{\partial^2\omega^1}{\partial\eta^2} - Re\xi\frac{\partial\omega^1}{\partial\xi} + Re(\eta+\epsilon)\frac{\partial\omega^1}{\partial\eta} = 0 \tag{6.77}$$

$$\frac{\partial^2\psi^1}{\partial\xi^2} + \frac{\partial^2\psi^1}{\partial\eta^2} + (\xi^2+\eta^2)\omega^1 = 0 \tag{6.78}$$

Solid Phase

$$\frac{\partial^2 T^1}{\partial\xi^2} + \frac{\partial^2 T^1}{\partial\eta^2} - P_c\xi\frac{\partial T^1}{\partial\xi} + P_c\eta\frac{\partial T^1}{\partial\eta} = 0 \tag{6.79}$$

Interface

$$\partial_\xi\psi^1 = \epsilon\left(\xi\eta_s^1\right)' \tag{6.80}$$

$$\partial_\eta\psi^1 = \epsilon\left(\eta_s^1\right)' \tag{6.81}$$

$$\left(\xi\eta_s^1\right)' + P_c(1+\epsilon)\eta_s^1 + P_c(1+\xi)^{-\frac{3}{2}} + \frac{\partial}{\partial\eta}\left(T_l^1 - T_s^1\right) = 0 \tag{6.82}$$

$$T_s^1 = T_l^1 - \eta_s^1 \tag{6.83}$$

$$T_s^1 = -(1+\xi)^{-\frac{3}{2}} \tag{6.84}$$

Infinity

$$T^1 \to 0 \tag{6.85}$$
$$\partial_\eta \psi^1 \to 0 \tag{6.86}$$
$$\partial_\xi \psi^1 \to 0 \tag{6.87}$$

at $\eta \to \infty$ as well as at $\xi \to \infty$.

Smooth Tip Condition

$$\left(\eta_s^1\right)'(0) = 0 \tag{6.88}$$
$$\eta_s^1(0) \quad = 1 \tag{6.89}$$

at $\xi = 0$.

Solving the First Order Equations

A solution of the first-order equations can be obtained via separation of variables [117]. The starting point of the following analysis is the **homogeneous heat conduction** equation associated with the inhomogeneous equation (6.75) valid in the liquid phase[7]:

$$\frac{\partial^2 T_h^1}{\partial \xi^2} + \frac{\partial^2 T_h^1}{\partial \eta^2} - P_c \xi \frac{\partial T_h^1}{\partial \xi} + P_c(\eta + \epsilon)\frac{\partial T_h^1}{\partial \eta} = 0 . \tag{6.90}$$

Based on the ansatz

$$T_h^1(\xi, \eta) = X(\xi)Y(\eta),$$

this can be rewritten as

$$X'' - P_c \xi X' + 2P_c \lambda X \qquad = 0 , \tag{6.91}$$
$$Y'' + P_c (\eta + \epsilon) Y' - 2P_c \lambda Y = 0 . \tag{6.92}$$

Choosing

$$x = \tfrac{P_c}{2}\xi^2, \ X = X(x) , \tag{6.93}$$

(6.91) is transformed into Kummer's equation

$$xX'' + \left(\frac{1}{2} - x\right) X' + \lambda X = 0 , \tag{6.94}$$

with a fundamental set of solutions given by

$$X(x) = \begin{cases} M(-\lambda, \tfrac{1}{2}, x) \ \text{(regular at } \xi = x = 0) \\ U(-\lambda, \tfrac{1}{2}, x) \ \text{(singular at } \xi = x = 0) . \end{cases} \tag{6.95}$$

---

[7] The definitions of special functions appearing in the succeeding analysis follow mainly [3], partly [219].

Here M and U are the confluent hypergeometric functions. Suppressing the divergence of the solution at $\xi = x = 0$, an appropriate choice of $X(x)$ is restricted to

$$X(x) = M\left(-\lambda, \frac{1}{2}, x\right) = 1 + \frac{(-\lambda)_1}{\left(\frac{1}{2}\right)_1}x + \frac{(-\lambda)_2}{\left(\frac{1}{2}\right)_2 2!}x^2 + \ldots + \frac{(-\lambda)_n}{\left(\frac{1}{2}\right)_n n!}x^n + \ldots , \quad (6.96)$$

where $(a)_n$ is the Pochhammer symbol, i.e. $(a)_n = a(a+1)(a+2)\ldots(a+n-1)$ and $(a)_o = 1$. On the other hand, $M(-\lambda, \frac{1}{2}, x)$ grows exponentially, i.e.:

$$M\left(-\lambda, \frac{1}{2}, x\right) \sim \frac{\Gamma\left(\frac{1}{2}\right)}{\Gamma(-\lambda)}x^{-\lambda-\frac{1}{2}}e^x \quad \text{as } x \to \infty , \quad (6.97)$$

unless $\Gamma(-\lambda)$ is infinite. This implies $\lambda = 0, 1, 2, \ldots$, in which case the series in (6.96) truncates. Thus the acceptable solutions of (6.91) are the Hermite polynomials

$$X(\xi) = M\left(-n, \frac{1}{2}, \frac{P_c}{2}\xi^2\right) = \frac{n!}{(2n)!}\left(-\frac{1}{2}\right)^{-n}H_{2n}\left(\sqrt{P_c}\xi\right) , \quad (6.98)$$

where $\lambda = n = 0, 1, 2, \ldots$

Moreover, turning towards (6.92) it is convenient to introduce

$$y = \sqrt{P_c}(\eta + \epsilon), \ Y = e^{-\frac{y^2}{4}}Z(y) \cdot \quad (6.99)$$

This results in Weber's equation:

$$Z'' - \left(\frac{y^2}{4} + 2n + \frac{1}{2}\right)Z = 0 , \quad (6.100)$$

which fundamental set of solutions read

$$Z(y) = \begin{cases} U\left(2n + \frac{1}{2}, y\right) = D_{-2n-1}(y) \\ V\left(2n + \frac{1}{2}, y\right) \end{cases} , \quad (6.101)$$

where U and V are parabolic cylinder functions. For large y they display the asymptotic behavior:

$$e^{-\frac{y^2}{4}}D_{-2n-1}(y) \sim x^{-2n-1}e^{-\frac{y^2}{2}}, \\ e^{-\frac{y^2}{4}}V\left(2n + \frac{1}{2}, y\right) \sim \sqrt{\frac{2}{\pi}}y^{2n} . \quad (6.102)$$

For exponential decay of the solution as $\eta \to \infty$, one has to choose $D_{-2n-1}(y)$ accordingly:

$$Y(\eta) = e^{-\frac{P_c}{4}(\eta+\epsilon)^2}D_{-2n-1}\left(\sqrt{P_c}(\eta + \epsilon)\right) . \quad (6.103)$$

Thus the solution of the homogeneous equation (6.90) can be expanded as

$$T_h^1(\xi,\eta) = \sum_{n=0}^{\infty} \alpha_n H_{2n}\left(\sqrt{P_c}\xi\right) \frac{e^{-\frac{P_c}{4}(\eta+\epsilon)^2} D_{-2n-1}\left(\sqrt{P_c}\,(\eta+\epsilon)\right)}{e^{-\frac{P_c}{4}(1+\epsilon)^2} D_{-2n-1}\left(\sqrt{P_c}\,(1+\epsilon)\right)} . \qquad (6.104)$$

As a consequence the overall solution of (6.75) for **inhomogeneous heat conduction** can be expressed based on a Greens function formalism as:

$$T_l^1(\xi,\eta) = T_h^1(\xi,\eta) - \int\limits_0^{\infty} d\xi' \int\limits_1^{\infty} d\eta' G(\xi,\eta;\xi',\eta') \frac{\partial \psi^1}{\partial \xi} \frac{\partial T^0}{\partial \eta}(\xi',\eta') . \qquad (6.105)$$

Here one can define the general formula for the Greens function $G(\xi,\eta;\xi',\eta')$ as a solution of

$$\frac{\partial^2 G}{\partial \xi^2} + \frac{\partial^2 G}{\partial \eta^2} - P_c\xi\frac{\partial G}{\partial \xi} + P_c(\eta+\epsilon)\frac{\partial G}{\partial \eta} = \delta(\xi-\xi')\delta(\eta-\eta') , \qquad (6.106)$$

with suitable boundary conditions at $\eta = \infty$ and $\xi = 0$. The Greens function can be obtained following the general outline given in [219]. As a function of ξ it can be expanded into a set of Hermite polynomials as given in (6.98). A general approach to G taking into account ξ- as well as η-dependence reads

$$G(\xi,\eta;\xi',\eta') = \sum_{n=0}^{\infty} A_n(\eta;\xi',\eta')H_{2n}(\sqrt{P_c}\xi) . \qquad (6.107)$$

Inserting (6.107) into (6.106) yields:

$$G(\xi,\eta;\xi',\eta') = \sum_{n=0}^{\infty} \left[\frac{\partial^2}{\partial\eta^2}A_n + P_c(\eta+\epsilon)\frac{\partial}{\partial\eta}A_n - 2P_c n A_n\right] H_{2n}(\sqrt{P_c}\xi)$$

$$= \delta(\xi-\xi')\delta(\eta-\eta') . \qquad (6.108)$$

At this point it is appropriate to take advantage of the orthogonality relation of the Hermite polynomials:

$$\int\limits_{-\infty}^{\infty} dt\, e^{-\frac{t^2}{2}} H_n(t)H_m(t) = \begin{cases} 0 & \text{for } n \neq m \\ \sqrt{2\pi n!} & \text{for } n = m \end{cases} . \qquad (6.109)$$

Since $H_{2n}(t)$ is an even function of t an alternative way to express this orthogonality is given by:

$$\int\limits_0^{\infty} d\xi\, e^{-\frac{P_c\xi^2}{2}} H_{2n}(\sqrt{P_c}\xi)H_{2m}(\sqrt{P_c}\xi) = \delta_{nm}\sqrt{\frac{\pi}{2P_c}}(2n)! . \qquad (6.110)$$

Applying the projection operator of (6.110) to (6.108) one obtains

$$\frac{\partial^2}{\partial\eta^2}B_n + P_c(\eta+\epsilon)\frac{\partial}{\partial\eta}B_n - 2P_c n B_n = \delta(\eta-\eta') , \qquad (6.111)$$

where

$$B_n = \frac{A_n \sqrt{\frac{\pi}{2P_c}}(2n)!}{e^{-\frac{P_c}{2}\xi'^2} H_{2n}(\sqrt{P_c}\xi')} .$$

Thus the problem is reduced to finding a one-dimensional Greens function defined by (6.111) with general form [3]:

$$B_n = \frac{1}{\Delta(y_1, y_2)} \begin{cases} y_1(\eta)y_2(\eta') \text{ for } \eta \le \eta' \\ y_2(\eta)y_1(\eta') \text{ for } \eta > \eta' \end{cases} . \qquad (6.112)$$

Here the Wronskian $\Delta(y_1, y_2)$ is evaluated at η'. y_1 and y_2 are two independent solutions of the homogeneous equation

$$\frac{\partial^2}{\partial \eta^2}y + P_c(\eta + \epsilon)\frac{\partial}{\partial \eta}y - 2P_c n y = 0 , \qquad (6.113)$$

as given by (6.99) and (6.101), or a suitable linear combination of both. The choice of y_1 and y_2 depends on the boundary conditions one imposes. Suppressing divergence at infinity results in:

$$y_2(\eta) = e^{-\frac{P_c}{4}(\eta+\epsilon)^2} D_{-2n-1}\left(\sqrt{P_c}\,(\eta+\epsilon)\right) . \qquad (6.114)$$

For y_1 one is more flexible and can introduce an additional parameter γ_n, which allows for a variety of boundary conditions one does not need to specify at this point. Thus:

$$y_1(\eta) = \left[\gamma_n D_{-2n-1}\left(\sqrt{P_c}\,(\eta+\epsilon)\right)\right.$$
$$\left. -V\left(2n+\frac{1}{2}, \sqrt{P_c}\,(\eta+\epsilon)\right)\right] e^{-\frac{P_c}{4}(\eta+\epsilon)^2} . \qquad (6.115)$$

For the subsequent calculation of the Wronskian $\Delta(y_1, y_2) = y_1 \frac{\partial}{\partial \eta}y_2 - y_2 \frac{\partial}{\partial \eta}y_1$ it is appropriate to define the abbreviations:

$$x = \sqrt{P}(\eta + \epsilon) \quad , \quad c = \gamma_n \quad , \quad u(x) = D_{-2n-1}\left(\sqrt{P_c}\,(\eta+\epsilon)\right) \quad ,$$
$$v(x) = V\left(2n + \frac{1}{2}, \sqrt{P_c}\,(\eta+\epsilon)\right) .$$

With these notations solutions y_1 and y_2 read:

$$y_1 = [cu(x) - v(x)]\,e^{-\frac{x^2}{4}} , \quad y_2 = u(x)e^{-\frac{x^2}{4}} .$$

Denoting differentiation with respect to x by a prime, the Wronskian can be rewritten as

$$\Delta(y_1, y_2) = y_1 \frac{\partial}{\partial \eta}y_2 - y_2 \frac{\partial}{\partial \eta}y_1 = \sqrt{P_c}(uv' - vu')e^{-x^2/2} = \sqrt{P_c}\Delta(u,v)e^{-x^2/2} ,$$

where $\Delta(u,v) = uv' - vu'$ is given in [3] as:

$$\Delta(u,v) = \sqrt{\frac{2}{\pi}} .$$

Thus:

$$\Delta(y_1, y_2) = \sqrt{\frac{2P_c}{\pi}} e^{-\frac{P}{2}(\eta'+\epsilon)^2} .$$

This is independent of $c = \gamma_n$. With these expressions the Greens function reads:

$$G(\xi, \eta; \xi', \eta') = e^{\frac{P_c}{2}[(\eta'+\epsilon)^2 - \xi'^2]} g(\xi, \eta; \xi', \eta') , \tag{6.116}$$

where

$$g(\xi, \eta; \xi', \eta') \tag{6.117}$$

$$= \sum_{n=0}^{\infty} \frac{1}{(2n)!} H_{2n}(\sqrt{P_c}\xi) H_{2n}(\sqrt{P_c}\xi') D_{-2n-1}\left(\sqrt{P_c}\,(\eta_> + \epsilon)\right) e^{-\frac{P_c}{4}(\eta_> + \epsilon)^2}$$

$$\cdot \left[\gamma_n D_{-2n-1}\left(\sqrt{P_c}\,(\eta_< + \epsilon)\right) - V\left(2n + \frac{1}{2}, \sqrt{P_c}\,(\eta_< + \epsilon)\right)\right] e^{-\frac{P_c}{4}(\eta_< + \epsilon)^2} .$$

Here $\eta_>$ refers to $\eta > \eta'$, whereas $\eta_<$ refers to $\eta < \eta'$.

To solve for **heat conduction in the solid phase**, separation of the two variables ξ and η is employed as well:

$$T_s^1(\xi, \eta) = X(\xi)Y(\eta).$$

This yields (6.91) and (6.92) with $\epsilon = 0$. Following the steps to derive (6.98), $X(\xi)$ is determined in the same manner as for the homogeneous heat conduction problem in the liquid, yielding (6.98). To solve (6.92) for $\epsilon = 0$ it is convenient to choose

$$y = -\frac{P}{2}\eta^2, \ Y = Y(y) . \tag{6.118}$$

Thus (6.92) for $\epsilon = 0$ is transformed into Kummer's equation (see (6.94)). The solution regular at $y = \eta = 0$ is $M(-n, \frac{1}{2}, y)$. As a consequence

$$Y(\eta) = M\left(-n, \frac{1}{2}, -\frac{P_c}{2}\eta^2\right) = \frac{n!}{(2n)!}\left(-\frac{1}{2}\right)^{-n} H_{2n}\left(i\sqrt{P_c}\eta\right) . \tag{6.119}$$

Therefore the solution of heat conduction in the solid can be expanded as:

$$T_s^1(\xi, \eta) = \sum_{n=0}^{\infty} \beta_n H_{2n}\left(\sqrt{P_c}\xi\right) \frac{H_{2n}\left(i\sqrt{P_c}\eta\right)}{H_{2n}\left(i\sqrt{P_c}\right)} . \tag{6.120}$$

Linear Solvability Theory

Accordingly y_1 of (6.115) is selected as

$$y_1(\eta) = \left[V\left(2n + \frac{1}{2}, \sqrt{P_c}\,(1+\epsilon)\right) D_{-2n-1}\left(\sqrt{P_c}\,(\eta+\epsilon)\right) \right.$$
$$\left. - D_{-2n-1}\left(\sqrt{P_c}\,(1+\epsilon)\right) V\left(2n + \frac{1}{2}, \sqrt{P_c}\,(\eta+\epsilon)\right) \right] e^{-\frac{P_c}{4}(\eta+\epsilon)^2} .$$

As a consequence γ_n is determined via (6.115) and thereby specifies the Greens function G based on (6.117). In turn G enters (6.105) and thus selects T_l', which has to satisfy the first-order boundary conditions (6.82)–(6.84). At this point one has to realize that it is impossible to satisfy all three of these boundary conditions together. Thus the steady state solutions derived in the beginning of this section are destroyed by isotropic surface tension.

Irregular Perturbation Expansion

However, from diffusion limited dendritic growth it is known as well that anisotropy of surface tension can cure the problem we encountered at the end of the last section, i.e. destruction of steady state solutions through isotropic surface tension. In fact, anisotropy of surface tension selects a discrete spectrum of growth velocities for the case of diffusion limited dendritic growth, of which only the fastest growth mode is linearly stable[8]. In this case it is necessary to study the additional effects arising from anisotropy in the complex plane, e.g. via a WKB formalism [236]. The key point of such an irregular perturbation expansion is the evaluation of a related eigenvalue problem in the complex plane. This eigenvalue problem itself arises from the difference between the governing equation with zero surface tension solution inserted respectively the perturbed solution inserted. Its analysis results in a matching condition between the regular and the irregular parts of the solution. One can analyze this solvability condition further to derive the selected tip velocity as:

$$V_{\text{Tip}} = \frac{D(1+\epsilon)\sigma(\beta)}{d_0} P_c^2(\Delta, \epsilon) , \qquad (6.121)$$

where $\sigma(\beta)$ refers to the dimensionless growth rate known from diffusion limited dendritic growth [42] and $P_c^2(\Delta, \epsilon)$ to the respective Peclet number, which here is a function of ϵ as well. The precise relation determining $P_c^2(\Delta, \epsilon)$ reads

$$\Delta = \sqrt{\frac{\pi P_c}{2}}\,(1+\epsilon) \exp\left[\frac{P_c}{2}(1+\epsilon)^2\right] \text{erfc}\left(\sqrt{\frac{P_c}{2}}(1+\epsilon)\right) . \qquad (6.122)$$

---

[8] A concise review article discussing the selection problem in the case of diffusion limited dendritic growth is given in [42].

It is obtained – just as in the case of diffusion limited dendritic growth – from the relation expressing the temperature for the zero surface tension steady state solution right at the interface. The $\Delta-$ as well as the $\epsilon-$dependence of P_c is visualized in Fig. 6.4.

6.3 Comparison to Experimental Data

Obviously it is $\epsilon = \frac{\rho_s - \rho_l}{\rho_l}$ which contains the density difference between liquid and solid phase and thus the origin of the additional driving force, which characterizes this growth problem compared to diffusion limited dendritic growth. Five different solutions for P_c corresponding to five different values of ϵ are plotted in Fig. 6.4.

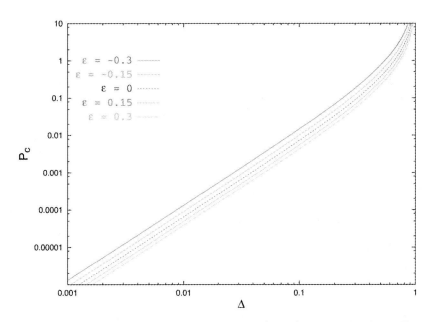

Fig. 6.4. The Peclet number P_c as obtained in (6.121) plotted for five different values of ϵ. The relation of these curves to experimental data is discussed in more detail in the text.

The case realized by most materials, namely $\epsilon > 0$, obviously results in reduced growth. For a real material such as e.g. succinonitrile with $\epsilon = 0.028$, at $\Delta = 0.01$, velocities are only 95.38% of the ones in the diffusion limited case and only 94.63% at $\Delta = 0.7$. The reason is that for $\rho_s > \rho_l$ density change flow drives a hydrodynamic flow towards the dendritic interface. This hinders the transport of latent heat away from the dendrite and thus slows down the solidification process.

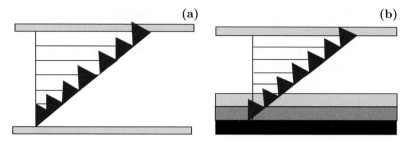

Fig. 6.5. Comparison of a no-slip condition for (a) sharp versus a no-slip condition for (b) diffuse interface models. The term forcing a hydrodynamic flow as depicted at the right enters as source term within the hydrodynamic equations of a diffuse interface model. It can be interpreted as a dissipative boundary force.

Moreover, the above results indicate that the effect of density change flow on a dendrite's tip evolution increases at low undercoolings. Again such a dependence on the undercooling can be explained by the fact that at low undercoolings the temperature gradient as main driving force becomes smaller and smaller. Thus it overrides hydrodynamic effects less and less. As a consequence one would expect to observe the effects of density change flow within these parameter regimes at least in microgravity experiments, where buoyancy does not override density change flow. Indeed – as indicated at the end of Sect. 6.1 – a difference between the theoretical solution for purely diffusion limited dendritic growth and data of microgravity experiments has been reported previously. One well known experiment along this line is the one by Glicksman *et al.* [129]. However, comparing the above theoretical results including the effect of density change flow to the experimental data by Glicksman *et al.* one realizes that the difference between theory and experiment is even larger than for diffusion limited dendritic growth theory (Fig. 6.2). This result is found in numerical simulations of respective phase-field model equations as well.

The precise model equations for the simulation results presented below are

Heat Conduction Equation

$$\frac{\partial T}{\partial t} + (\mathbf{w}\cdot\nabla)\, T = \nabla(D_T\,\nabla T) - \frac{\partial \Gamma_L(\phi)}{\partial t}$$

Phase – Field Equation

$$\tau\frac{\partial \phi}{\partial t} + (\mathbf{w}\cdot\nabla)\,\phi = \xi^2\nabla^2\phi + V_0(\phi - \phi^3) + M_0\frac{1 - \phi^2}{(1 + \phi^2)^2}(T - 1 + \Gamma_L(\phi))$$

Navier – Stokes Equation

$$\frac{\partial \mathbf{w}}{\partial t} + (\mathbf{w}\cdot\nabla)\,\mathbf{w} = -\frac{1}{\rho_l}\nabla p + \nu\nabla^2\mathbf{w} + \Gamma_{NS}$$

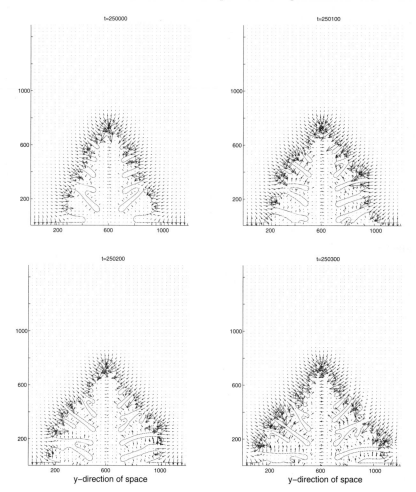

Fig. 6.6. Simulation depicting four different stages of density change influenced growth for $\epsilon = 0.3$. The evolution runs from upper left to upper right, lower left and lower right. The corresponding time steps of the numerical simulation are displayed above each figure. Further details are given in the text.

Incompressibility Condition

$$\nabla \cdot \mathbf{w} = 0 \quad \text{with } \Gamma_{NS} = \frac{1}{2}\gamma\,\kappa\,\nabla\phi$$

$$\text{and } \Gamma_L(\phi) = \frac{\phi}{1 + \phi^2} \ .$$

Diffusion advection type equations are implemented on the basis of an ADI (alternating direction implicit) scheme [133]. The Navier–Stokes type equations are solved employing a MAC (Marker and Cell) solver with discretization of space coordinates as proposed in [137]. All simulations are done on a quadratic grid. Numerical parameters for the simulation runs are $\xi = 0.9$, $V_0 = 0.4$, $\tau = D_T = 1$, $Re = 1500$, $\Delta t = 0.5$, $\Delta x = 0.3$, $N_x = 1500$ and $N_y = 1200$, where N_x and N_y denote the grid size in $x-$ and $y-$direction respectively. The initial condition is a parabolic interface of height 75 grid units with the hydrodynamic velocity field identical to zero in the solid phase. Nevertheless after a few time steps hydrodynamic flow is encountered within the solid phase and clearly visible in the graphs. In the literature it is known as *parasite current*. It arises due to the fact that in the phase-field simulation the no-slip condition gets smeared over the diffuse interface region (Fig. 6.5). These parasite currents have been investigated in more detail by Lafaurie *et al.* [195]. Applying their results allows me to choose appropriate parameter regimes, where the effect of these parasite currents remain bounded and quantitative changes to the simulation results are negligible. The side branching activity in dendritic growth as visible in the simulation pictures is known to result from thermal fluctuations [22, 44, 180, 201, 234]. These have to be included in the simulations by transforming

$$T_{i,j} \rightarrow T_{i,j} + R_0(r_{i,j} - \frac{r_{i+1,j} + r_{i-1,j} + r_{i,j+1} + r_{i,j-1}}{4}) \qquad (6.123)$$

with noise amplitude R_0 and random numbers $r_{i,j}$ evenly distributed in the interval $[-\frac{1}{2}; \frac{1}{2}]$. Taking into account the $r_{i\pm1,j\pm1}$ as in (6.123) ensures local energy conservation. The noise amplitude R_0 corresponds to the magnitude of thermal noise encountered in dendritic growth experiments, which reads (in dimensional units):

$$\frac{k_B T_M^2 c_p}{L^2 d_0^2} . \qquad (6.124)$$

This is equal to the mean square fluctuation of T inside a two-dimensional volume of d_0^2. For a comparison Fig. 6.6, Fig. 6.7 and Fig. 6.9 depict dendritic evolution at different driving forces and noise levels, i.e. $M_0 = 4$, $M_0 = 2.5$ and $M_0 = 1$, respectively. For $M_0 = 2.5$ the evolution of the temperature field is displayed, as well, through colored contour plots in Fig 6.8. Simulations were carried out keeping the position of the dendritic tip at approximatly same height by repeated transformations of the x-coordinate (coordinate in growth direction) of all field variables. In any case the growth rate of the dendritic tip is smaller than in the case of corresponding simulations without coupling to the Navier–Stokes equations resulting in an increased deviation from the experimental values by Glicksman *et al.*. Thus at this point one clearly has to draw the conclusion that even taking into account surface tension corrections within the energy condition, density change flow cannot explain the difference between experiment and theory.

Other physical mechanisms active in microgravity experiments must have an even stronger influence of reversed sign. Studying the literature one will

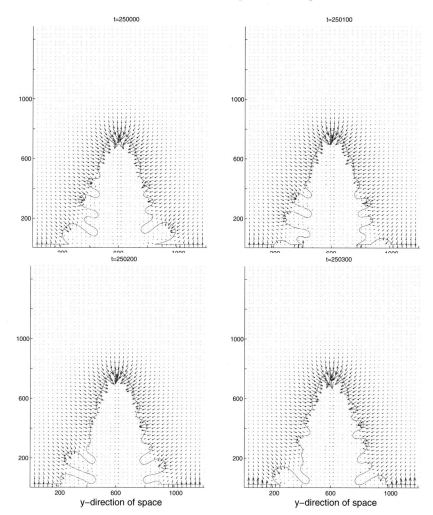

Fig. 6.7. Four different stages of growth demonstrating the effect of a reduced driving force with $M_0 = 2.5$ and otherwise same parameters as in Fig. 6.6.

find a couple of articles discussing possible mechanisms, as e.g. finite-size effects through container walls [83] or nutation of the dendrite under microgravity [251]. Meanwhile, the experimentalists themselves called the experimental conditions within the space shuttle a *reduced gravity* environment (see e.g. [194]) thus hinting at the possibility of some "left-over buoyancy" convection influencing growth at low undercoolings. To my knowledge quantitative studies of the possible effects of "left-over buoyancy" convection have not been reported. Rather the recent focus of work by Glicksman *et al.* is concerned with fluctuations of the dendritic tip velocity [84, 193, 194]. However,

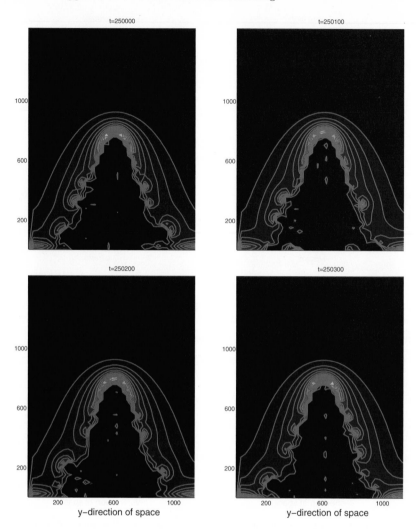

Fig. 6.8. The evolution of the temperature field corresponding to the four different stages of growth depicted in Fig. 6.7.

an explanation of the systematic deviation between diffusion theory and the experimental curve for $\Delta \leq 0.3\,K$ receiving community wide acceptance has not yet been given.

In this context a phenomenon, which to my knowledge has never been studied in detail even though it has been assumed to influence dendritic growth at low undercoolings quantitatively [32], is the influence of thermo-capillary convection. It arises from a weak temperature dependence of surface tension[248]. In dendritic growth the temperature along the interface is not constant. Rather - taking into account a fourfold crystalline anisotropy - there

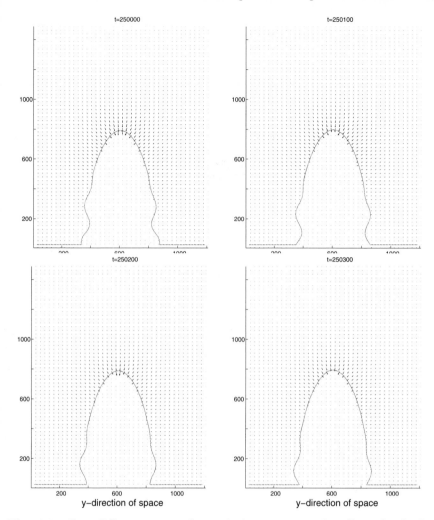

Fig. 6.9. Four different stages of growth demonstrating the effect of a further reduction of driving force at $M_0 = 1.0$ (otherwise same parameters as in Fig. 6.6).

are two coldest points to the left and to the right of the tip position as indicated in Fig. 6.10. Thus there is a temperature gradient along the dendritic interface from the tip towards these points. This drives a thermocapillary flow as depicted in Fig. 6.10. Comparing its direction to the one of flow driven by density change (Fig. 6.6), one finds it to be reversed. Thus also its quantitative effect on the dendritic growth velocity should be reversed, i.e. it should enhance growth compared to diffusion limited dendritic growth. In that sense it could be an explanation for the experimental data in [129].

Fig. 6.10. The two coldest points of the parabolic interface are to the left and to the right of the tip as indicated by the dark dots. If surface tension is temperature dependent, the respective temperature gradient comes along with a surface tension gradient, which is the origin of thermocapillary convection.

The problem is to fully grasp the quantitative effects of thermocapillary flow on the evolution of a dendritic tip. From the point of view of sharp interface modeling one would include it as a balance condition for pressure, viscous flow stresses and surface tension:

$$(p_s - p_l) \cdot \mathbf{n} + \eta(\nabla \mathbf{u} + (\nabla \mathbf{u})^T) \cdot \mathbf{n} - \nabla\gamma = 0 , \qquad (6.125)$$

where $\nabla \mathbf{u} + (\nabla \mathbf{u})^T$ represents the viscous stress tensor for a liquid obeying the incompressible Navier–Stokes equations. The surface tension term $\nabla\gamma$ can be expanded in a Taylor series truncated after its first term, i.e.:

$$\nabla\gamma = \frac{\partial\gamma}{\partial T}\nabla T . \qquad (6.126)$$

Here $\frac{\partial\gamma}{\partial T}$ is a material parameter, which is usually less than 0. Equation (6.125) yields a condition for the gradients of the flow velocity field normal to the interface. It occurs as an additional boundary condition at the interface, which has to be solved together with (6.30) and (6.31).

Trying to apply a perturbative analysis as sketched in section 6.2 to model equations (6.26)–(6.38) with additional boundary condition (6.125) one will find that a solution, for which vorticity vanishes everywhere, is no longer compatible with all boundary conditions. It is not obvious how to construct an ansatz solving the Navier–Stokes equations with boundary conditions (6.30), (6.31) *and* (6.125), which could serve as starting point for the subsequent perturbative analysis.

A numerical investigation would be an alternative. However, in this case it is not clear, whether the phase-field method with its smearing of boundary conditions over the diffuse interface could handle the problem with necessary precision. In particular the findings of Lafaurie *et al.* [195] with respect to parameter regimes, in which effects of parasite currents remain bounded, cannot be carried over directly.

Preliminary results without thorough investigation of stability and convergence of the algorithm are depicted in Fig. 6.11 just to demonstrate qualitatively the effect of thermocapillary convection at a dendritic interface in

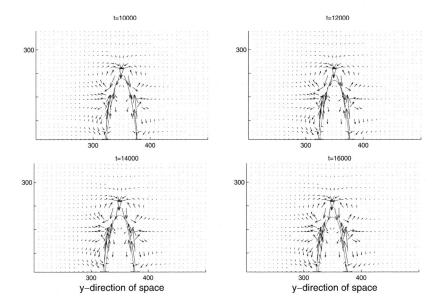

Fig. 6.11. Simulation depicting the influence of thermocapillary convection at a dendritic interface. Calculation were carried out at $\Delta = 0.03$ assuming quasi-incompressibility and setting $\eta(\nabla \mathbf{u} + (\nabla \mathbf{u})^T) \cdot \mathbf{n} = \nabla \gamma$ within the interfacial region (no noise included). Judged based on the velocity components depicted here the dynamics of fluid flow seems to be restricted to a small layer around the dendritic interface. However the pressure field displays a distribution which reaches far ahead of the dendritic tip (see Fig. 6.11).

simulations. They are obtained by assuming quasi-incompressibility and setting $\eta(\nabla \mathbf{u} + (\nabla \mathbf{u})^T) \cdot \mathbf{n} = \nabla \gamma$ within the interfacial region. For the material parameter $\frac{\partial \gamma}{\partial T} = -0.028 \frac{\text{N}}{\text{mK}}$ these preliminary numerical investigations indicate an increase of v_n of as much as 14.7 % at $\Delta = 0.03$, clearly demonstrating the effect required to explain the experiments[9]. A more thorough investigation is in preparation [111].

6.4 Summary

Within this chapter a phase-field model for hydrodynamically influenced dendritic growth was employed to derive the sharp interface equations governing

---

[9] This value $\frac{\partial \gamma}{\partial T}$ is an estimate, since to my knowledge experimental measurements of the precise value of $\frac{\partial \gamma}{\partial T}$ for SCN are not yet available. However, it has been measured for silicon as $\frac{\partial \gamma}{\partial T} = -0.28 \frac{\text{N}}{\text{mK}}$ [144]. On the other hand, the surface tension itself is usually an order of magnitude smaller for organic materials than for silicon [24]. Thus to get an estimate for $\frac{\partial \gamma}{\partial T}$ applying to SCN I scaled the value obtained for silicon with the ratio of surface tensions of the two materials. The idea to proceed like this is due to Glicksman [131].

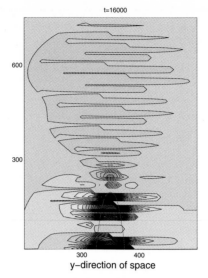

Fig. 6.12. The pressure field around a dendrite growing influenced by thermo-capillary convection. This figure corresponds to the final graph in the evolution of Fig. 6.11.

density change flow. Compared to the classical approach towards dendritic growth the resulting model equations contain an additional curvature term in the energy boundary condition. The effects of such a term had been investigated by Lemieux *et al.* for diffusion limited dendritic growth [205]. In that case it results in an increase of selected velocities. In that sense including it into an analysis of density change influenced dendritic growth was accompanied by the motivation to find an explanation for the experimental data in [129]. The analysis itself was carried out as proposed originally by Kruskal and Segur [191]. In the context of dendritic growth this method was employed by various authors [236]. It starts from the derivation of a zero surface tension solution. An analysis of this solution within the framework of regular perturbation expansion reveals that isotropic surface tension destroys the steady state solutions found without surface tension. An irregular perturbation expansion allows me to analyze the problem taking into account crystalline anisotropy. The selected velocity thereby obtained is (6.122). Its difference to the velocity selected in diffusion limited dendritic growth is contained in the terms depending on ϵ, the density change between liquid and solid normalized with respect to ρ_l. Effects of the additional curvature term obtained via the asymptotic analysis of Sect. 6.1 are of higher order in the expansion. Thus other than for pure diffusion limited dendritic growth, where exactly this curvature term gives rise to increased values of selected velocities [205], within the mathematical treatment underlying (6.90)–(6.121) it cannot be employed to explain larger growth velocities than for diffusion limited dendritic growth. Thus the contradiction to the experimental data obtained in the experiments of [129] cannot be solved at this point. To explain the latter experiments other possible influences as indicated in Sect. 6.3 remain to be investigated and understood quantitatively.

6.5 Appendix

Appendix A: Transformation to Parabolic Coordinates

Interface

$$s(\xi, \eta, t) = \eta - \eta_s(\xi, t) = 0 \tag{6.127}$$

Unit Normal and Tangent Vector

$$\nabla s = \frac{2}{\sqrt{\eta + \xi}} \left(-\sqrt{\xi} \eta_s' e_\xi + \sqrt{\eta} e_\eta \right) \tag{6.128}$$

$$|\nabla s| = \frac{2}{\sqrt{\eta + \xi}} \sqrt{\xi (\eta_s')^2 + \eta} \tag{6.129}$$

$$\mathbf{n} = \frac{\nabla s}{|\nabla s|} = \frac{1}{\sqrt{\xi (\eta_s')^2 + \eta}} \left(-\sqrt{\xi} \, \eta_s' e_\xi + \sqrt{\eta} \, e_\eta \right) \tag{6.130}$$

$$\mathbf{t} = \frac{1}{\sqrt{\xi (\eta_s')^2 + \eta}} \left(\sqrt{\eta} \, e_\xi + \sqrt{\xi} \, \eta_s' \, e_\eta \right) \tag{6.131}$$

Normal Velocity

$$\mathbf{v} \cdot \mathbf{n} = -\frac{\frac{\partial s}{\partial t}}{|\nabla s|} \tag{6.132}$$

Curvature

$$x_s(\xi) = \sqrt{\xi} \eta_s(\xi) \tag{6.133}$$

$$y_s(\xi) = \frac{1}{2} \left(\eta_s(\xi) - \xi \right) \tag{6.134}$$

$$\kappa[\eta_s(\xi)] = \frac{y_s' x_s'' - x_s' y_s''}{\left((y_s')^2 + (x_s')^2 \right)^{\frac{3}{2}}} = \frac{(\eta_s - \xi \eta_s')^2 (1 - \eta_s') - 2\xi (\xi + \eta_s) \eta_s \eta_s''}{(\xi + \eta_s)^{\frac{3}{2}} (\eta_s + \xi (\eta_s')^2)^{\frac{3}{2}}} \tag{6.135}$$

Anisotropy

$$\cos\theta = \mathbf{n} \cdot \mathbf{e_x} = \frac{\xi\eta'_s + \eta_s}{\sqrt{\eta_s + \xi}\sqrt{\xi(\eta'_s)^2 + \eta_s}} \tag{6.136}$$

$$\sin\theta = \mathbf{t} \cdot \mathbf{e_x} = \frac{\sqrt{\xi\eta}\,(n'_s - 1)}{\sqrt{\eta_s + \xi}\sqrt{\xi(\eta'_s)^2 + \eta_s}} \tag{6.137}$$

$$\cos(4\theta) = -8\sin^2\theta\cos^2\theta + 1 = -8\frac{\xi\eta_s(\eta'_s - 1)(\xi\eta'_s + \eta_s)}{(\xi + \eta_s)^2\,(\xi(\eta'_s)^2 + \eta_s)^2} + 1 \tag{6.138}$$

$$A[\eta_s(\xi)] = 1 - \beta + \frac{8\beta\xi\eta_s(\eta'_s - 1)(\xi\eta'_s + \eta_s)}{(\xi + \eta_s)^2\,(\xi(\eta'_s)^2 + \eta_s)^2} \tag{6.139}$$

7. Application to Epitaxial Growth Involving Elasticity

In Sect. 5.2.3 the phenomenon of epitaxial surface growth has been introduced as an application of an asymptotic analysis in the thin interface limit. It was demonstrated, how a set of phase-field model equations can be developed from two thermodynamic potentials, namely the inner energy functional $\mathcal{E}$ and the entropy functional $\mathcal{S}$. Thereby an additional degree of freedom is obtained. This additional degree of freedom allows me to recover the precise attachment kinetics of the two-sided epitaxial growth problem involving unequal attachment kinetics and discontinuities at the interface asymptotically at first order (*thin-interface limit*). The sharp interface equations constituting this two-sided epitaxial growth problem are given by (5.68)–(5.71).[1] Moreover, the corresponding phase-field model equations are (5.80) and (5.81). For further details the reader is referred to Sect. 5.2.3.

The scope of the succeeding two sections is to exploit the variational priciples of irreversible thermodynamics to extend the above phase-field model equations to elastic effects at strained surfaces [110]. In this manner the phenomenon of epitaxial growth at strained surfaces becomes numerically tractable. On the other hand, trying to extend the above sharp interface equations to elastic driving forces results in boundary conditions, which in their general formulation cannot be solved, as will be demonstrated in more detail in Sect. 7.1. This possibility to obtain tractable models where the sharp interface approach fails can be viewed as one of the advantages of phase-field modeling justifying the development of this method. I will come back to this point in Chap. 8.

---

[1] Note that in the case of epitaxial growth anisotropy due to the underlying crystal structure of the material is contained in $\rho^{eq} = \rho_0^{eq} \cdot (1 - \beta cos 4\theta)$. Detailed studies of the influence of crystalline anisotropy in epitaxial growth can be found in [106, 109].

7.1 Elastic Driving Forces Within the Diffuse Interface Approach: Extended Model Equations for Strained Surfaces

If an epitaxial surface consists of strained layers, the correct free energy to describe the system has to include the respective elastic strain energy. Thus (5.79) changes into:

$$f(\phi, \rho, u_{ij}) = \frac{V_E + \rho V_S}{2} d(\phi) + \frac{(\rho - \rho^{eq})}{\rho^{eq}} k(\phi) - \rho \ln(\frac{\rho}{\rho^{eq}}) + \rho + f(\phi)_{\text{el}} . \quad (7.1)$$

Here f_{el} is a free energy density, which takes into account the global elastic energy in both phases. A natural starting point to construct a suitable form of f_{el} for the phase-field approach is the relation between the stress tensor σ_{ij} and the strain tensor u_{ij} as given by Hooke's law:

$$f = \mu\, u_{ij} u_{ij} + \frac{\kappa}{2} u_{ii}^2 , \quad (7.2)$$

where summation over double subscripts is implied. κ and μ are the Lamé constants. For plane strain, these elastic constants are related to Young's modulus E and the Poisson ratio v via $\mu = E/[2(1+v)]$ and $\kappa = Ev/[(1+v)(1-2v)]$. f the free energy per unit volume of a general system, of which the only energy contribution is elasticity. The stress tensor σ_{ij} is then given by:

$$\sigma_{ij} = \frac{\partial f}{\partial u_{ij}} = 2\mu u_{ij} + \kappa u_{kk}\delta_{ij} = 2\mu(u_{ij} - u_{kk}\frac{\delta_{ij}}{d}) + K u_{kk}\delta_{ij} , \quad (7.3)$$

where $K = \kappa + 2\mu/d$ is the bulk modulus and d referring to the dimension for which the problem is solved (i.e. $d = 2$ or $d = 3$). If there is a misfit on the surface of the substrate, additional misfit strain arises. Following [261] this additional strain can be modeled through an additional contribution σ_{ij}^m:

$$\sigma_{ij}^m = -2\mu\epsilon^m \left[\frac{1+v}{1-2v}\delta_{ij} \right] , \quad (7.4)$$

where ϵ^m quantifies the misfit. It is then straightforward to formulate the following ansatz for the ϕ-dependence of f:

$$f_{\text{el}}(\phi, \{u_{ij}\}) = k(\phi) \left\{ \mu u_{ij} u_{ij} + \frac{\kappa}{2} u_{ii}^2 - 2\mu\epsilon^m \left[\frac{1+v}{1-2v} \right] \right\}$$
$$+ [1 - k(\phi)] \frac{\tilde{\kappa}}{2} u_{ii}^2 . \quad (7.5)$$

Here $\tilde{\kappa}$ is the bulk modulus of the non-solid phase, i.e. liquid or vapor depending on the employed epitaxial technique. As a result the governing equations of the phase-field model given in an adiabatic approximation, i.e. $\delta F_{el}/\delta u_i = 0$ (for implications of this approximation see the paragraph following (7.7)), now read:

$$\tau \frac{\partial \phi}{\partial t} = (\xi + \rho \xi)\nabla^2 \phi - \frac{V_E + \rho V_S}{2} d'(\phi) - \frac{\rho - \rho^{eq}}{\rho^{eq}} k'(\phi)$$

$$+ k'(\phi)[\mu u_{ij} u_{ij} + \frac{\kappa - \tilde{\kappa}}{2}(u_{ii})^2 - 2\mu \epsilon^m \left[\frac{1 + v}{1 - 2v}\right]], \qquad (7.6)$$

$$\frac{\partial \rho}{\partial t} = D\Omega \nabla^2 \rho + \mathcal{G}(\phi)\frac{\partial \phi}{\partial t} + F + \frac{\rho}{\tau} . \qquad (7.7)$$

These have to be solved together with the equations for the evolution of the elastic variables, where the strains u_{ij} $(i, j = 1, ..., d)$ are dependent quantities. Therefore, the variational derivatives $\delta F/\delta u_{ij}$ are dependent, as well. To proceed one may exploit the fact that the components u_i $(i, j = 1, ..., d)$ of the displacements are independent variables. Assuming that the relevant time scales of the problem are large compared to sound propagation times, the variational derivatives $\delta F_{el}/\delta u_i$ of the elastic free energy can be assumed to be zero, which is exactly the adiabaticity assumption. Hence one obtains

$$0 = \frac{\delta F_{el}}{\delta u_i} = \frac{\partial}{\partial x_j}\frac{\delta F_{el}}{\delta u_{ij}} = \frac{\partial}{\partial x_j}\{k(\phi)\sigma_{ij} - [1 - k(\phi)]\kappa u_{kk}\delta_{ij}\} . \qquad (7.8)$$

The second term can be related to the pressure of the system via

$$p = p_{0\ell} - \tilde{\kappa} u_{kk} , \qquad (7.9)$$

where $p_{0\ell}$ has been chosen as equilibrium pressure in the liquid phase. This term contributes only in the liquid phase. Assuming that during the experiment on the liquid side deviations from this equilibrium pressure remain small, (7.8) can be simplified as follows:

$$0 = \frac{\partial}{\partial x_j}\left\{k(\phi)\sigma_{ij}\right\} . \qquad (7.10)$$

Equations (7.6), (7.7) and (7.10) become closed equations by replacing σ_{ij} and u_{ij} by the field variables u_i using the definition of the strain tensor,

$$u_{ij} = \frac{1}{2}\left(\frac{\partial u_i}{\partial x_j} + \frac{\partial u_j}{\partial x_i}\right) , \qquad (7.11)$$

and Hooke's law. Its applicability is demonstrated in the following section. Again an ADI scheme is employed for the diffusion-advection type equations (7.6) and (7.7). The elastic problem, on the other hand, is solved via successive overrelaxation, where the time integration is performed by a formally second-order accurate midpoint scheme [220].

Turning towards the sharp interface model given by (5.68)–(5.71) accounting for elastic effects requires a coupling to the following evolution equations for the displacements of the solid phase:

$$(1 - 2v)\partial_k^2 u_i + \partial_i \partial_k u_k = 0 \ . \tag{7.12}$$

To evaluate these equations, boundary conditions of the displacement field at the moving, phase-separating interface have to be determined at every time step of the calculation. The correct physical boundary condition is a balance of the force acting on the film surface and the pressure in the liquid (vapor) phase, i.e.:

$$\sigma_{ij} \cdot n_j = p^{l/v}|_{\Gamma_{ij}} \ . \tag{7.13}$$

Obviously in this case the construction of the correct boundary values for the displacements u_{ij} based on (7.13) is less straightforward than the formulation of the phase-field equations (7.6)–(7.11). A more detailed discussion of this point can be found for the special case of an assumed zero-pressure condition above the substrate in [261].

7.2 Comparing Simulations and Experiments

For strained surfaces grown by epitaxy two qualitatively different morphologies are known to result from the epitaxial process, namely

1. rippling of continuous surface layers,
2. a morphology made up of an array of islands.

Based on the experimental observations reported in [6, 86, 87] ripples were associated with layers of a low misfit surface. On the other hand, islands were understood to be the dominating morphology in high misfit regimes [94, 97, 204, 218]. In contrast to this, Dorsch *et al.* [95] found a morphology transition towards islands even for low misfit surfaces. Their experiments were done for strained $Ge_x Si_{1-x}$ grown from In solution at low driving forces. Dorsch *et al.* claimed that their findings coincide with theoretical work by Gao [123], indicating, that in all misfit regimes island morphologies are the ones to result in maximum strain energy relaxation. Following that point of view one would expect islands to dominate the morphological surface evolution independent of the misfit, once growth follows essentially energetics. Accordingly epitaxial growth at strained surfaces could be viewed as a two stage process:

1. During a first stage of growth a rippled morphology with well defined initial wavelengths arises.
2. At a subsequent stage of growth evolution is dominated more and more by the principle of maximum strain energy relaxation, which drives the system into a transition towards an island morphology. This transition is

independent of the surface miscut. It is an expression of the rippled morphology itself, the geometry of which results in a redistribution of strain. In turn this changes the surface potential such that it favors the strain relaxation mechanism. During this stage of growth the island morphology essentially inherits the wavelength of the initial ripple instability.

Thus a transition to islands should occur independent of the misfit at advanced stages of growth. A necessary presupposition is diffusion in the nourishing phase (liquid or vapor, depending on the epitaxial technique) as material transport mechanism. Previous to the experiments by Dorsch et al. a ripple morphology was usually observed resulting from MBE (molecular beam epitaxy) experiments. Compared to LPE (liquid phase epitaxy) the diffusion coefficients encountered in MBE are four orders of magnitude smaller. This leads to the conjecture that during MBE diffusion processes are simply not fast enough to trigger the morphology transition towards islands before the end of the experiment. In that sense the ripple morphologies observed in MBE experiments would have to be considered transiental morphologies. Moreover, pseudomorphic islands were the "true" stationary growth mode to be expected from any kind of epitaxial setting at a strained surface.

From point of view of a nonlinear analysis carried out for strained epitaxial surfaces by Spencer et al. [261] this is not necessarily the case. Their analysis revealed two different steady state solutions within the near critical parameter regime:

solution 1: a spatially periodic small amplitude cusp like solution
solution 2: a spatially periodic large amplitude solution of sinusoidal
           shape.

Both solutions were found to be unstable, leaving open the question, whether the evolution of the strained surface would be steady outside the realm of long-wave theory or characterized by a transient state displaying coarsening.

To reconcile these findings with the experimental observations described above, one would have to identify the island morphology as the "steady state solution outside the realm of long-wave theory", which develops eventually as secondary instability from solution 1 as well as from solution 2. However, the question remains: Which of the two qualitatively different types of primary solutions 1 or 2 are the ripples observed initially in the experiments by Dorsch et al.? Are they "spatially periodic finite amplitude rounded cusp solutions" or rather "near critical spatially periodic small amplitude solutions"? If either of them is true, might the morphology observed in MBE experiments actually be the other kind of solution? And if so, might the secondary instability of MBE growth be the second kind of scenario depicted above, namely a transient state displaying coarsening? If this were true, it could still be reasonable not to observe the coarsening process due to slow diffusion as already pointed out by Albrecht et al. [95]. Nevertheless the precise understanding of the dynamics of strained surface epitaxial growth would be different - a

difference which might become important to predict the correct morphology
for different parameter settings. Thinking in this direction one could for example wonder, whether high driving forces and/or high temperatures might
result in a transition towards islands even for MBE growth.

Table 7.1. Amplitudes and cell widths for varying diffusivities D, $\epsilon^m = 0.05$

D	1	10^{-2}	10^{-4}	10^{-6}
Amplitude	73	59	45	13
Cell width	6	5	4	1

To answer the questions above a full simulation of model equations (7.6)–
(7.11) is a valuable tool. Note that for the simulations reported in this chapter
the terms $F + \frac{\rho}{\tau}$ (see (7.7)) are discarded. This refers to the situation of no
desorption, which is true for MBE experiments as well as for LPE. Moreover,
to model LPE growth as employed in the experiments by Dorsch *et al.*, one
has to keep in mind that the experimental conditions are chosen such that the
solutal field is approximately at equilibrium. The true driving force is rather
the temperature field T, which takes the place of the variable ρ in (7.6) and
(7.7). This exchange of ρ and T is possible, since under the assumption of
local thermodynamic equilibrium the phase diagram allows us to relate the
two to each other. Approximating solidus line and liquidus line of the diagram
as straight lines (see Fig. 2.4) the relevant relations read:

$$C^S = -\left|\frac{dC^S}{dT}\right|(T - T_m) \tag{7.14}$$

$$C^L = -\left|\frac{dC^L}{dT}\right|(T - T_m) = -\frac{1}{m_l}(T - T_m) , \tag{7.15}$$

where C^S is the concentration of solute in the solid phase, C^L its concentration in the liquid phase, T_m the melting temperature and m_l the slope of
the liquidus line, which is usually negative. In this way the model allows us
to investigate the evolution of strained MBE as well as LPE surface growth
taking into account all of the nonlinear modes and thus to identify "steady
state solutions outside the realm of long-wave theory".

Table 7.2. Amplitudes and cell widths for varying misfits ϵ^m, $D = 1$

ϵ^m	0.05	0.01	0.12	0.15
Amplitude	73	34	21	9
Cell width	6	3	2	1

In particular it enables an investigation of the amplitude and the shape of the evolving morphology (sinusoidal versus cusp like) depending on the diffusion constant as well as the misfit angle. The results are summarized in the tables. In Table 7.1 the misfit ϵ^m is kept fixed, whereas D varies. In Table 7.2, on the other hand, D is kept fixed and ϵ^m varies. The cell widths reported in the two tables are measured at one half of the amplitude height. The remaining parameters of our investigations read $\Delta x = \Delta y = 0.005$, $\Delta t = 0.0018$, $\epsilon^2 = 1$, $\Psi = 1$, $\gamma = 0.7417$, $\nu = 0.0021$, $\kappa = 1.4371$, $\upsilon = 0.3212$, $\kappa = 0.6666$, $\tilde{\kappa} = 1$ and $\mu = 0.3341$. Again all parameters are scaled with respect to Δx and Δt. Note that within the context of this work we identify a morphology to be cusp like if its first derivative does not display a turning point between two extrema[2] (Fig. 7.1). Obviously, the higher the misfit ϵ^m the lower the amplitude of the primary instability. This coincides with the experimental findings of Dorsch et al. just as well as with theoretical work by Srolovitz [262]. The diffusion constant D displays the same qualitative behavior. Moreover, the unproportionally large drop in amplitude for D below 10^{-4} indicates that the transition to a different type of solution occurs. Indeed the small amplitude solutions found by simulation of (7.6)–(7.10) for $D = 10^{-6}$ display a cusp like shape with considerably smaller widths of the cell shaped instabilities. Thereby they differ from the solutions for $D = 1$ to $D = 10^{-4}$. Examples for both solutions are given in Fig. 7.1.

Taking into account that low diffusion constants belong to MBE growth, whereas large diffusion constants indicate use of the LPE technique, one important result of this numerical investigation is that for equal misfits the primary ripple like instabilities of MBE and LPE differ: first, amplitudes of LPE instabilities are larger; second, their tips are rounder. On the other hand, tips of MBE instabilities display a cusp like shape. The question which remains is whether these different solutions display differences with respect to the long time evolution as well. Figure 7.1 presents a view on the temporal evolution of the respective solutions. Four subsequent stages of growth are given with intervals of 10^4 time steps in between. In Fig. 7.1b a transition of ripples to islands indicated by negative growth values in the valleys of the morphology becomes visible. In contrast valleys of the cusp like instabilities keep positive growth rate. After $t = 6 \cdot 10^4$ time steps their shape is approximately stationary. Signs of coarsening cannot be detected. These numerical findings are indeed characteristic for the long time behavior of cusp like solutions versus large amplitude solutions, i.e. they were obtained for other sets

---

[2] A fully cusped tip could not be resolved via phase-field simulations, for which the interface width l puts an inherent length scale cut off. l has to be resolved by 8-10 grid points to avoid numerical metastability [38]. The results reported here were obtained for four different values of l with different mesh sizes to fulfill the above criterion. They are independent of this change of numerical parameters. Thus in the investigated parameter regimes no fully cusped solutions can be expected.

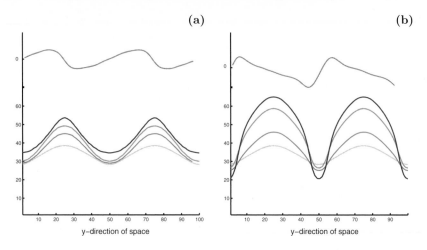

Fig. 7.1. Contour plots of surface morphologies at $t = 3 \cdot 10^4$ (turquois line), $t = 4 \cdot 10^4$ (red line), $t = 5 \cdot 10^4$ (green line) and $t = 6 \cdot 10^4$ (blue line) for $\epsilon = 0.15$: (a) evolution of the cusp like morphology, $D = 10^{-6}$, (b) large amplitude morphology, $D = 1$. The pink lines depict the local slopes of the contours at $t = 6 \cdot 10^4$ (blue line). The derivatives are generated from the data points of the contour plots. Note that they are aligned with respect to a second zero point on the vertical axis.

of diffusion constants and misfits, as well. To elucidate these findings further, Fig. 7.2 displays the evolution of the principal component of the strain tensor in growth direction. The different strain fields arise as an answer to the different evolutions of the two respective surfaces and in turn influence their further growth as well. I.e., surface regions subject to low strain have high growth rates, whereas the regions exposed to large strain exhibit slow growth or even recede. In this sense in Fig. 7.1b enhanced amplitude growth of the morphology can be attributed to the large difference of strain values around its crest versus the one around its valleys. For the cusp like solutions, on the other hand, strain is distributed more evenly over the surface.

Together these numerical studies indicate that even though MBE and LPE experiments at strained surfaces can both result in similar rippled surface morphologies, these ripples are not exactly the same type of solution of the growth problem. While the ripples found for low diffusion constants, i.e. MBE growth conditions, are cusp like instabilities, the ripples found in LPE are large amplitude solutions with rounded tips. In the long time evolution the latter will undergo a transition towards islands. Cusp like solutions do not display such a transition even after simulation of time intervals much longer than any experiment. However, there is no sign of coarsening either. The latter finding can be compared to recent numerical studies of a strained surface in contact with its melt, but not subject of epitaxy [172]. Results of these studies indicate that a primary instability of periodic grooves will undergo

(a)   **(b)**

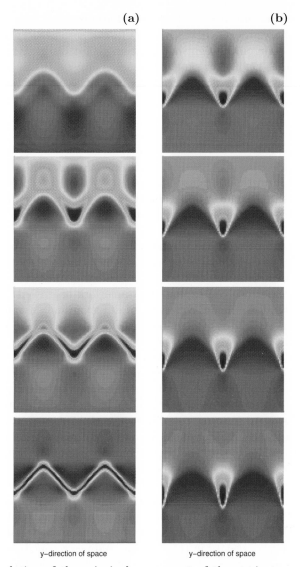

y–direction of space   y–direction of space

Fig. 7.2. Evolution of the principal component of the strain tensor in growth direction: (a) cusp like morphology, (b) large amplitude morphology. The time steps, for which the fields are displayed, coincide with the ones for the contour plots of Fig. 7.1, i.e. from top to bottom: $t = 4.5 \cdot 10^4$, $t = 5 \cdot 10^4$, $t = 5.5 \cdot 10^4$ and $t = 6 \cdot 10^4$.

coarsening only in the absence of gravity. Once gravity is taken into account, several stationary grooves coexist within the final surface morphology. Within (7.6) the last term on the right-hand side modeling the surface miscut could be interpreted as having the same influence as gravity. In that sense it would stabilize the cusp like rippled morphology in a way, such that coarsening does not occur.

7.3 Discussion

Here epitaxial growth at strained surfaces was chosen as a second example to demonstrate the convenient applicability of the phase-field approach when it comes to the point of including additional driving forces into an interfacial growth problem. Since such additional driving forces can very easily be related to an additional term in the system's free energy density functional, they directly enter the variational principles on which the derivation of phase-field model equations is based upon.

In this manner one can exploit the framework of irreversible thermodynamics to extend the model equations of diffusion limited epitaxial growth as derived in Sect. 5.2 to stress induced surface effects. The respective model equations have been developed in detail in Sect. 7.1.

In Sect. 7.2 they have been applied to experiments by Dorsch *et al.* [95]. In this context their simulation sheds new light on the morphology transition from ripples to islands. In particular the numerical investigations reveal that the ripple instability occurring in MBE has to be distinguished from the one in LPE growth. It seems to fall into the class of cusp like instabilities. The non-appearance of a morphology transition towards islands in principle hints at the possibility that the true secondary instability belonging to this kind of cusp like solution is a transiental coarsening regime. The reason that the coarsening is depressed, can be explained by comparison to the numerical studies in [172], where the authors report the absence of coarsening in a similar situation, if gravity effects are taken into account. It seems plausible that within the model equations derived here, the misfit term has the same stabilizing effect as gravity in [172]. These findings reconcile with the interpretation given by Dorsch *et al.* [95] in the sense that the formal reason for the morphological differences of MBE compared to LPE instabilities is given by the large ratio of the respective diffusion constants. As a consequence a morphology of cusp like ripples should be the one to be expected in MBE experiments even at high driving forces and high temperatures after extended deposition time – a point, which certainly could be clarified by experiments. If it were true, MBE would generally fail as a technique to tailor regular island arrays with specified island distance as functional devices.

8. Conclusions and Perspectives

Concluding the text at this point can only result in a partial conclusion: Research in the field of diffuse interface modeling is still young and fully active. It is likely, that while I am writing, further progress of considerable impact is achieved. As discussed progress could basically be twofold, either concerning modeling related issues or numerical ones.

With respect to the latter the numerical appendix of this book summarizes the most recent computational approach, which achieved outstanding computational efficiency for the case of 3D dendritic growth involving fluid flow [155] on the basis of a semi implicit approximated projection method, implemented fully adaptive and parallelized. It is an example of a growth phenomenon, which certainly cannot be treated numerically without employing enhanced techniques. However, for this kind of moving boundary problems an implementation making use of enhanced numerical techniques is possible only based on diffuse interface modeling. The reason is – as pointed out in Chap. 2 – that only the diffuse interface approach allows for an Euler scheme[1]. Thus one conclusion to draw at this point is, that a diffuse interface approach for moving boundary problems becomes necessary, if one intends to apply the full power of high performance computing, which today is essentially connected to parallelization, to their simulation.

With respect to modeling-related progress, the concept of thermodynamic consistency was described in detail in Chaps. 3–5 as a most important underlying viewpoint. It was explained, how a thermodynamically consistent diffuse interface model can be obtained by the variation of underlying thermodynamic potentials. Thermodynamic consistency itself can be interpreted as a validation of the model. Consequently one can go on to examine the derived model equations with the intention to understand more about the behavior of physical fields in the region of the moving, phase separating interface. The mathematical method allowing for the respective analysis, is the

---

[1] As discussed in Chap. 2 a level set approach would basically yield an alternative for the implementation of a moving boundary employing an Euler scheme, as well. However, from point of view of numerics such a level set approach imposes difficulties a diffuse interface approach does not (see Chap. 2). Moreover the level set approach is lacking the thermodynamic background of diffuse interface models and thus qualifies only as a numerical tool for classically known moving boundary problems.

method of matched asymptotic expansion described in Chap. 5. Applications of this concept of diffuse interface modeling given in the text are twofold: The first is devoted to hydrodynamically driven dendritic growth (Chap. 6). This example was chosen to demonstrate, how the asymptotic analysis of a diffuse interface model can be employed to revise a traditional sharp interface formulation of a moving boundary problem. Subsequent analysis of the specified model equations via a perturbation expansion around a zero surface tension steady state solution yields new insight in the dynamics of dendrites growing at low undercoolings. In particular it sheds new light on the well known and still not fully explained experiments [129], in which Glicksman *et al.* carefully determined the difference between dendritic tip velocities in microgravity experiments and diffusion limited growth theory. In this context it is the achievement of the diffuse interface approach to allow for a rigorous derivation of a specified set of model equations revising the relation between experiment and theory.

A second example was presented in Chap. 7 to illustrate the usefulness of diffuse interface modeling with respect to a different class of moving boundary problems: As demonstrated for the case of elastically influenced epitaxial surface growth, sharp interface formulations of moving boundary problems involving more complex transport than just diffusive one might yield boundary conditions, which numerically are no longer tractable. Here the diffuse interface approach offers the great advantage of overcoming the need to solve boundary conditions at the phase separating interface. Thus for this type of problems one can conclude, that to find a numerical solution at all, it is unavoidable to turn towards diffuse interface modeling.

These examples combined with the theoretical background given in the first five chapters are intended to provide the reader with an idea of what diffuse interface modeling is all about, for what classes of problems it is useful and what kind of knowledge it allows us to derive based on a combination of thermodynamic modeling, mathematical analysis and simulation. If interested in the field for the purpose of an own application, a first step would be to formulate the underlying free energy or entropy functional as basis for the variational approach, by which the relevant model equations are to be obtained (Chap. 4). Today's applications range from dendritic growth to electrodeposition, from image processing to superconducting materials, from alloy solidification to polymer systems. Certainly depending on the focus of one's interest the complications when applying diffuse interface modeling to any of these phenomena are more but just one step beyond diffusion limited dendritic growth, from which I started my discussion taking it as paradigmatic problem in Chap. 2. Therefore it seems appropriate to reflect for a moment on what really has been achieved in the field of diffuse interface modeling and what problems one would judge to be in reach based on the knowledge already gained.

In the following I will restrict the discussion of what has been achieved to the phenomenon of crystallization from the melt, which is directly linked to the paradigmatic dendritic growth problem. Thus it provides a representative overview of what one has to meet when turning from pure diffusion limited growth to further complications originating from additional transport fields[2], or from anisotropic kinetics and energetics in the interfacial region.

In the context of crystallization diffuse interface modeling refers back to Cahn [56]. His approach was taken up again several years later by Halperin et al. [143], Langer [200], Fix [120] and Collins and Levine [74], which then started a rapid development of diffuse interface modeling for crystal growth phenomena. Caginalp [49, 50] and others demonstrated the relation of these approaches to classical sharp interface models by means of asymptotics and distinguished limits. For sufficiently thin diffuse interfaces the Gibbs–Thomson boundary condition was proven to be incorporated in the diffuse interface approaches [52]. First derivations from an entropy functional were due to Penrose and Fife [231]. This came along with an awareness of the question of thermodynamic consistency of such models. In this context Wang et al. [289] introduced the concept of local positive entropy production and non-classical fluxes. At the same time Kobayashi [186, 187] demonstrated the relevance of diffuse interface modeling for real computations by simulating rather complicated solidification patterns. Since then many others [182, 239, 240, 290, 297] have used the diffuse interface approach successfully to study solidification morphologies. Great advance in the field was achieved through the introduction of the thin interface limit by Karma and Rappel [168]. It lessened the constraints with respect to the thickness of the interfacial region and thus rendered models computationally more efficient. The authors themselves used the increased capability of diffuse interface modeling gained from the thin interface limit amongst other things to study side branching effects in dendritic growth [168, 169]. Moreover the diffuse interface approach could be extended to study alloy solidification, i.e. a growth problem with coupled transport fields. Respective models were formulated by Wheeler et al. [295, 296], Bi et al. [252] and Charach and Fife [67, 68]. These could be employed to study the phenomenon of solute trapping [69, 76, 77, 296], dependence of surface tension on composition [69], isothermal dendritic growth and microsegregation [75, 291], Ostwald ripening and coalescence [292], recoalescence during dendritic solidification [35], cell to front phase transitions during directional solidification [36] and eutectic alloy solidification [96, 100, 101, 138, 166, 298]. A further issue was to blend hydrodynamics with diffuse interface modeling. Such models have been developed [26, 51, 272] and used to some extent in computation [85, 108]. Moreover it was shown, that with an additional potential modeling van der Waals and polarization forces, the models can be employed to simulate wetting phenomena, as well [221]. Only recently thermodynamically consistent

---

[2] which might not be continuous at the interface

diffuse interface models including hydrodynamics could be formulated for pure materials as well as for binary alloys by Anderson *et al.* [12, 13]. The same authors also developed a model including convection and anisotropy for a pure material [15, 17]. This model was subsequently extended by Sekerka and Bi [250] to multicomponent alloy solidification. In the area of solid–solid transformation Braun *et al.* [39, 40] have used a diffuse interface field model with multiple order parameters to explore order–disorder phase boundaries and transformation in alloys. Moreover, Kobayashi *et al.* [188] have developed a phase-field model involving two order parameters, expressed in terms of polar coordinates, that facilitate modeling of polycrystalline materials with grain boundaries. Finally Karma [171] was able to solve the problem of two-sided dendritic growth. Thus a wide range of algorithmic means to tackle complexities arising from coupled transport fields as well as kinetic and energetic anisotropies in the interfacial region are available from this research in the field of crystallization by now.

Turning from the discussion of "What has been achieved?" towards the question of "What can be achieved?" it seems natural to point out, that there are still questions to investigate for the phenomena of electrodeposition, of image processing, of polymer systems and so on, where one would simply have to carry over steps taken in the field of crystallization from the melt. However, there is a further complexity tied to some of these phenomena, namely an inherent multi-scale nature of the problem due to an internal structure of the material. Examples for such materials with inner structure are liquid crystals and polymers. For these kind of systems the diffuse interface approach opens an additional interesting perspective, since it can be used as basis of a rigorous mathematical multi-scale approach. Thus the sentence of the introductory chapter "One might wonder if in the end this approach can provide a framework to tackle the behavior of still more complicated systems, e.g. systems with an inherent multi-scale nature due to an internal structure, such as liquid crystals or polymer solutions, as well." is actually an anticipated perspective at this point. There is a clear methodological framework to tackle the multi-scale problem, namely *homogenization*. Examples of how homogenization can be applied to diffuse interface modeling are [209] and [98]. Further steps to treat phase transitions in materials with inner microstructure in a thermodynamically consistent manner are provided by Kalospiros *et al.* [159, 160] in the field of polymer growth. Now entering the Greek culture allows me to close these remarks unscientifically by turning towards Plato: "And with respect to the other points I would not dare to adhere to my view with great insistence, but ... I ... believe that we must try to find what is not known, for we should be ... less idle than if we believed that what we do not know it is impossible to find out and that we need not even try."

Plato – words from the mouth of Socrates, Menon 86C [243]

A. Numerical Issues
of Diffuse Interface Modeling

This appendix is concerned with the questions: What happens after a diffuse interface model has been established? How straightforward is its implementation? Is there anything one has to take special care of with respect to that implementation? The answer to the latter is a clear *yes*. Answering it in more detail will leave the question of the previous sections whether or not a diffuse interface model under consideration is thermodynamically consistent behind. Rather one is concerned with a choice of model parameters which will yield the best numerical performance. For dendritic growth it could be proven that from point of view of run time performance models which do not display thermodynamic consistency and can only be validated by the method of matched asymptotics expansion can be superior to thermodynamic consistent ones [182]. In that sense referring to the implementation point of view one finds once more[1] the expression *universality* of diffuse interface models in the previous citation.

Since the early attempts of phase-field computation in the context of solidification by Smith [259] great advances in the accuracy of those computations have been encountered. The reasons are twofold: First, computer technology has advanced significantly and thus allows us to integrate the unsteady phase-field equations for increasingly complex configurations. Second, there has also been progress in the numerical techniques applied to the implementation of phase-field models. This involves choosing an optimal set of numerical parameters and model potentials just as well as considering enhanced numerical techniques as parallel, adaptive or multi-grid algorithms.

The first of the above two issues arises since

1. different potentials can be employed, resulting in different phase-field models, which nevertheless describe the same growth phenomenon in their sharp interface limit and
2. to establish the correspondence between the diffuse interface model and the sharp interface model there is some freedom in how to arrange the numerical parameters of the diffuse interface, i.e. a physical variable ap-

---

[1] "once more" refers to the discussion at the end of Chap. 5, where the notion of *universality* in the context of phase-field modeling has been used to denote a freedom in the precise choice of $\mathcal{F}$.

pearing in the sharp interface model is usually expressed by more than one numerical parameter of the diffuse interface model.

I will demonstrate the above in detail in Sect. A.1 and Sect. A.2 for the case of dendritic growth, for which it has received much attention by the scientific community over the last decade, e.g. [113, 182, 232]. Whereas Sect. A.1 is devoted to the relationship between physical variables and numerical parameters, thereby pointing out the freedom in the precise choice of model parameters, Sect. A.2 compares the computational efficiency of different phase-field models depending on the choice of the underlying themodynamic potential.

In Sect. A.3 I will proceed to discuss some state of the art numerical approaches to interfacial growth phenomena and thereby give a look at how advanced numerical techniques can be employed to speed up diffuse interface computation – a step which is certainly required to deal with more complex, challenging growth conditions.

A.1 Relationship Between Physical Variables and Numerical Parameters

With respect to phase-field computations in dendritic solidification Kobayashi [187] was the first to become famous for quite beautiful animations of 3D phase-field simulations of evolving dendrites. However, this work was still very qualitative. To use phase-field simulation for quantitative predictions of dendritic growth morphologies one has to take a closer look at the relationship between the physical variables of the sharp interface limit and the numerical parameters of the phase-field model equations.

For diffusion limited dendritic growth the Gibbs–Thomson relation, the Stefan condition and the diffusion equation for the transport of heat in the two bulk phases constitute the free boundary problem, which describes the evolution of the dendritic front. The respective model equations (2.5)–(2.7) are rewritten here by transforming to a dimensionless temperature T via

$$T = \frac{c_p(T' - T'_\infty)}{L} \, .$$

(A.1)

This transformation employs the specific heat of the solid given by c_p. Moreover, T'_∞ refers to the dimensional temperature of the undercooled melt far from the interface. In addition one can introduce a dimensionless supercooling

$$\Delta = \frac{c_p(T'_M - T'_\infty)}{L}$$

(A.2)

of the liquid phase which is normalized, so that for $\Delta = 1$ the solidification of one volume element of the liquid phase results in precisely the heat required to heat up the same volume element from T'_∞ to T'_M. With these transformations the set of dimensionless equations corresponding to (2.5)–(2.7) reads:

$$\frac{\partial T}{\partial t} = D\nabla^2 T \tag{A.3}$$

$$T = \Delta - d_0(\theta)\kappa \tag{A.4}$$

$$\mathbf{v}\cdot\mathbf{n} = D\left[(\nabla T)_s - (\nabla T)_l\right]\cdot\mathbf{n} . \tag{A.5}$$

For large driving forces the consideration of local thermodynamic equilibrium is not valid any longer. This is taken into account by introducing a kinetic term $\beta_{\text{kin}}\cdot\mathbf{v_n}$ into the Gibbs–Thomson relation and thus replacing the dimensionless equation (A.4) by:

$$T = \Delta - d_0(\theta)\kappa - \beta_{\text{kin}}\mathbf{v}\cdot\mathbf{n} . \tag{A.6}$$

Here β_{kin} denotes a kinetic coefficient, which is a material parameter (compare to (2.10)). $d_0(\theta)$ is the capillary length given by (2.12).

For quantitative application of a derived phase-field model to a concrete dendritic growth problem, the capillary length d_0 and the kinetic coefficient β_{kin} of (A.6) have to be related to the numerical parameters of the phase-field model. To demonstrate how this can be achieved, I introduce a concrete phase-field model for dendritic growth of a binary alloy by specifying $G(\Phi, \mathbf{X}, \mathbf{r})$ in the remainder of this section. Following [2] I choose

$$G(\Phi, Q, C) = \frac{\xi^2}{2}(\nabla\Phi)^2 + V_0\left(\frac{\Phi^4}{4} - \frac{\Phi^2}{2}\right) \tag{A.7}$$
$$+ M_0\left\{\frac{C^2}{2} - C\frac{Q^2}{2} + Q + 1 + \Gamma_L(\Phi)(1 - C - Q)\right\} ,$$

with

$$\Gamma_L \equiv \frac{1}{2} + \frac{\Phi}{1 + \Phi^2} . \tag{A.8}$$

Here C denotes the solutal field. To overcome the problems related to phase-field modeling based on the dimensionless temperature T (see Sect. 4.2) the dimensionless energy density Q is used instead. It is related to T by

$$Q = \begin{cases} T & : \quad \text{solid phase} \\ T+1 & : \quad \text{liquid phase} . \end{cases} \tag{A.9}$$

With this choice of G and after retransformation from Q to T the set of phase-field model equations I discuss here reads

$$\tau\frac{\partial\Phi}{\partial t} = \xi^2\nabla^2\Phi + V_0(\Phi - \Phi^3) + M_0\frac{1 - \Phi^2}{(1 + \Phi^2)^2}\left(T + \frac{\Phi}{1 + \Phi}\right) , \tag{A.10}$$

$$\frac{\partial C}{\partial t} = \nabla\left(D_C\nabla(C - \Gamma_L(\Phi))\right), \tag{A.11}$$

$$\frac{\partial T}{\partial t} = \nabla(D_T\nabla T) - \frac{\partial\Gamma_L(\Phi)}{\partial t} . \tag{A.12}$$

In (A.10)–(A.12), it is the set of parameters ξ, V_0, τ and M_0 for which relations to the capillary length d_0 and the kinetic coefficient β_{kin} have to be found. To do so it is convenient to assume that d_0 and β_{kin} are constants, which are independent of the temperature field at the interface [61]. Moreover the Tolman–Buff effect [226], which describes a correction of the surface tension originating from curvature, is neglected. Then it is justified to derive the respective dependencies for a planar front as follows:

Since the capillary length d_0 is proportional to the surface tension γ, d_0 is obtained by comparing the Gibbs free energy $\mathcal{G}_{SIM}$ of an infinite, one-dimensional system with sharp interface to the Gibbs free energy $\mathcal{G}_{PFM}$ of the corresponding phase-field model with finite interface width. Assuming the interface at $x = 0$, it is the limit $\xi \to 0$, i.e. $\Phi(x) = \pm 1$ for $x \gtrless 0$, in which $\mathcal{G}_{SIM}$ follows from $\mathcal{G}_{PFM}$:

$$\mathcal{G}_{PFM} = \int_{-\infty}^{\infty} \frac{\xi^2}{2}\left(\frac{\partial\Phi(x)}{\partial x}\right)^2 + V_0\left(\frac{\Phi(x)^4}{4} - \frac{\Phi(x)^2}{2}\right) \tag{A.13}$$
$$+ M_0\left(\frac{C(x)^2}{2} - C(x) + \frac{Q(x)^2}{2} + Q(x) + 1 + \Gamma_L(\Phi)(1 - C - Q)\right) dx$$

$$\mathcal{G}_{SIM} = M_0 \int_{-\infty}^{\infty}\left(\frac{C_S^2}{2} + \frac{Q_S^2}{2} - C_S + Q_S + 1\right) dx . \tag{A.14}$$

$C_{L,S}$ and $Q_{L,S}$ are thermodynamic equilibrium values of C and Q in liquid and solid phase, respectively. Since the surface energy γ is defined as the difference of $\mathcal{G}_{PFM}$ and $\mathcal{G}_{SIM}$, one obtains an expression for γ by inserting the stationary solution of the phase-field model, which is given by

$$x = \sqrt{2}\xi \int_0^{\Phi(x)} \frac{1+y^2}{\sqrt{V_0(y^4-1)^2 + M_0(y^2-1)^2}}dy , \tag{A.15}$$
$$C(x) = C_0 + \Gamma_L(\Phi(x)) \quad \text{and} \tag{A.16}$$
$$Q(x) = Q_0 + \Gamma_L(\Phi(x)) . \tag{A.17}$$

Thus

$$\gamma = \int_{-\infty}^{\infty} \frac{\xi^2}{2}\left(\frac{\partial\Phi(x)}{\partial x}\right)^2 + V_0\left(\frac{\Phi(x)^4}{4} - \frac{\Phi(x)^2}{2}\right) + M_0\left(\frac{1}{2} - \Gamma_L(\Phi(x))^2\right) dx$$
$$= \sqrt{2}\xi\sqrt{M_0}\int_0^1 \frac{\sqrt{\frac{V_0}{M_0}(y^4-1)^2 + (y^2-1)^2}}{1+y^2}dy . \tag{A.18}$$

Equation (A.18) relates the surface energy to the phase-field parameters ξ, V_0 and M_0. From this the ξ, V_0 and M_0–dependence of d_0 can be obtained as

$$d_0 = \frac{\sqrt{2}\xi}{\sqrt{M_0}}\int_0^1 \frac{\sqrt{\frac{V_0}{M_0}(y^4-1)^2 + (y^2-1)^2}}{1+y^2}dy . \tag{A.19}$$

A more detailed derivation based on a generalized Gibbs–Thomson relation, which starts from the specification of liquid and solid chemical potentials with respect to an idealized phase diagram, is given in [2].

To derive an analogous expression for β_{kin}, I will turn to the quasistationary solutions of the phase-field model equations. For a velocity $v \ll 1$ they solve the differential equations

$$-\tau v \frac{\partial \Phi}{\partial x} = \xi^2 \frac{\partial^2 \Phi}{\partial x^2}$$

$$+ V_0(\Phi - \Phi^3) + M_0 \frac{1 - \Phi^2}{(1 + \Phi^2)^2} \left(C + T + 2\frac{\Phi}{1 + \Phi^2} \right) \quad \text{(A.20)}$$

$$-v \frac{\partial T}{\partial x} = D_T \frac{\partial^2 T}{\partial x^2} + v \frac{\partial \Gamma_L(\phi(x))}{\partial x} \quad \text{(A.21)}$$

$$-v \frac{\partial C}{\partial x} = D_C \frac{\partial^2 C}{\partial x^2} + v \frac{\partial \Gamma_L(\phi(x))}{\partial x} \ , \quad \text{(A.22)}$$

with stationary solutions ($v = 0$) (A.15) and

$$T_0(x) = -C_0(x) \ = \ \text{const} . \quad \text{(A.23)}$$

If one considers the physically relevant sharp interface limit, for which the diffusion lengths $l_D = \frac{2D_T}{v}$ and $l_C = \frac{2D_C}{v}$ are large compared to the diffuse interface width, in (A.21) and (A.22) the terms with factor v can be neglected and C as well as T can be assumed to be constant in the framework of an adiabatic approximation: $T(x) = T_0$; $C(x) = C_0$. From the phase-field equation (A.20) one obtains the following expression after integration of Φ in the interval $[-1; 1]$:

$$T_0 + C_0 = \frac{\sqrt{2}\tau v}{\sqrt{M_0}\xi} \int_0^1 \frac{\sqrt{\frac{V_0}{M_0}(y^4 - 1)^2 + (y^2 - 1)^2}}{1 + y^2} dy . \quad \text{(A.24)}$$

Since $T_{\mathrm{Int}} + C_{\mathrm{Int}} = T_0 + C_0 = -\beta_{\mathrm{kin}}v$, this can be rewritten to yield

$$\beta_{\mathrm{kin}} = \frac{\sqrt{2}\tau}{\sqrt{M_0}\xi} \int_0^1 \frac{\sqrt{\frac{V_0}{M_0}(y^4 - 1)^2 + (y^2 - 1)^2}}{1 + y^2} dy . \quad \text{(A.25)}$$

Moreover because of (A.19) one obtains a relation between the capillary length and the kinetic coefficient, which reads

$$\beta_{\mathrm{kin}} = \frac{d_0 \tau}{\xi^2} . \quad \text{(A.26)}$$

Thus in contrast to the sharp interface model the phase-field model displays model inherent kinetic effects [192]. Therefore, carrying out quantitative phase-field simulations for materials with vanishing kinetic effects requires to work in parameter regimes, for which the effect of the kinetic term $\beta_{\text{kin}} v_n$ is small compared to the curvature term $d_0 \kappa$. However, for those parameter regimes phase-field modeling is usually inefficient due to the large diffusion lengths originating from the low driving forces.

The application of phase-field models to dendritic growth of vanishing kinetic effects at moderate to large velocities became possible only after the introduction of the *isothermal* approach proposed by Karma and Rappel [167], as mentioned already in Chap. 1 and Chap. 5. The basic idea underlying that approach is to take explicitly into account the dependence of the temperature and solutal fields on the space variables within the interfacial region $|x| \lesssim \xi$. Essentially this implies carrying out the asymptotic analysis described in Sect. 5.1 up to the first order. This yields a solvability condition, by which the expression for the kinetic coefficient β_{kin} can be obtained. For the solidification of a binary alloy it reads:

$$\beta_{\text{kin}} = \frac{\sqrt{2}\tau I_3}{\sqrt{M_0}\xi} \left(1 - \frac{\xi^2}{\tau I_3} \left(\frac{1}{D_T} + \frac{1}{D_C} \right) \left(\frac{1}{2} I_1 + 2 I_2 \right) \right) . \tag{A.27}$$

In (A.27) the abbreviations I_1, I_2 and I_3 denote integrals, the specific form if which depends on the precise version of the phase-field model equations. For (A.10)–(A.12) they are of the form

$$I_1 = \int_0^1 \frac{(y-1)^2}{\sqrt{\frac{V_0}{M_0}(y^4-1)^2 + (y^2-1)^2}} dy \tag{A.28}$$

$$I_2 = \int_0^1 \frac{1-y^2}{(1+y^2)^2} \int_0^y \frac{\tilde{y}}{\sqrt{\frac{V_0}{M_0}(\tilde{y}^4-1)^2 + (\tilde{y}^2-1)^2}} d\tilde{y} dy \tag{A.29}$$

$$I_3 = \int_0^1 \frac{\sqrt{\frac{V_0}{M_0}(y^4-1)^2 + (y^2-1)^2}}{1+y^2}) dy . \tag{A.30}$$

The prefactor $\frac{\sqrt{2}\tau I_3}{\sqrt{M_0}\xi}$ in (A.27) corresponds precisely to (A.25), i.e. the sharp interface limit of the kinetic coefficient derived neglecting the spatial changes of the temperature and solutal field within the interfacial region. Thus in (A.27) the expression in parentheses denotes a correction term originating from the consideration of non-constant behavior of the two diffusive fields in the region $|x| \lesssim \xi$. For

$$\frac{\sqrt{M_0}\xi^2}{\tau I_3} \left(\frac{1}{D_T} + \frac{1}{D_C} \right) \overset{(A.19)}{=} \frac{\xi^3}{\tau d_0} \left(\frac{1}{D_T} + \frac{1}{D_C} \right) \to 0 \tag{A.31}$$

the limit of the correction term is unity, which is the condition for the applicability of the sharp interface limit as given by [168]. Thus in this limit (A.27) is equal to (A.25).

The great importance of (A.27) results from the fact that it allows us to minimize the correction term by enlarging ξ or reducing V_0/M_0. Finally for

$$\frac{\xi^2(D_T + D_C)}{\tau D_T D_C} = \frac{I_3}{\frac{1}{2}I_1 + 2I_2} \quad \text{or} \quad \frac{d_0^2 M_0(D_T + D_C)}{2\tau D_T D_C} = \frac{I_3^3}{\frac{1}{2}I_1 + 2I_2} \quad \text{(A.32)}$$

it vanishes. Equation (A.32) provides a condition, which allows us to carry out quantitative phase-field computations for the solidification of binary alloys with vanishing kinetic coefficient. Together with (A.19) and (A.25) it constitutes the set of expressions relating numerical phase-field parameters to physical parameters.

As mentioned above (A.28)–(A.30) and as a result (A.32), as well, depend on the specific choice of phase-field model equations. Therefore within the context of the phase-field model given by (A.10)–(A.12) a necessary condition for $\beta_{\rm kin} \to 0$ is $\xi > \sqrt{\frac{2\tau D_T D_C}{D_T + D_C}}$. For other models this is not necessarily the case. Some of them allow for a vanishing kinetic coefficient simply by variation of M_0 [167].

A.2 Computational Universality of Phase Field Models

Here I will turn from discussing a freedom in the precise choice of phase-field parameters towards a freedom with respect to the precise choice of the potential underlying the model equations by referring to the notion of *universality* of phase-field models. The first time the term universality of phase-field models appeared within this text, it was used exactly to denote such a freedom with respect to the choice of the free energy $\mathcal{F}$ underlying the derivation of a diffuse interface model. In this sense it refers to the fact that a large class of free energies give rise to the same sharp interface equations. Speaking of *computational universality* exploits this fact and goes a step further: If there is a freedom in the choice of the precise potential and if there is also some freedom in the choice of numerical parameters of a phase-field model as described in the previous section, then how does this manifest itself in the computational efficiency of a phase-field model? Are CPU times and convergence rates in the end factors to distinguish between different models and classify them? The discussion in this section following [182] will reveal that the latter is not the case. In this sense it will establish the notion of *computational universality*.

In the remainder of this section time-dependent solutions of different phase-field models for dendritic solidification in two dimensions are compared. To this end extensive computations using a specially developed adaptive mesh refinement algorithm [239, 240] are reported. These indicate that when properly used, all phase-field models give precise results, i.e. not only does each phase-field model converge to the steady state predicted by theory, but also the transient dynamics approach the steady state uniquely. Indeed, once one has established that there is genuine universal dynamic behavior, the only remaining question is that of computational efficiency. The results discussed here clearly indicate that the CPU times required for the different models are identical. In particular, there is no advantage for thermodynamically consistent models. Moreover finite discrepancies of the interfacial Peclet number Pe_{Int} encountered between different models are shown to be eliminable by adjusting the phase-field parameters.

The phase-field equations considered here are of the form

$$\frac{\partial T}{\partial t} = D\nabla^2 T + \frac{1}{2}\frac{\partial h(\phi)}{\partial t} \tag{A.33}$$

$$\tau(\mathbf{n})\frac{\partial \phi}{\partial t} = \nabla \cdot (W^2(\mathbf{n})\nabla\phi) - \frac{F(\phi, \lambda T)}{\partial \phi} + \frac{\partial}{\partial x}\left(|\nabla\phi|^2 W(\mathbf{n})\frac{\partial W(\mathbf{n})}{\partial \phi_x}\right)$$

$$+\frac{\partial}{\partial y}\left(|\nabla\phi|^2 W(\mathbf{n})\frac{\partial W(\mathbf{n})}{\partial \phi_y}\right) , \tag{A.34}$$

as in [168, 169]. Here the order parameter is denoted by ϕ, with $\phi = +1$ in the solid and $\phi = -1$ in the liquid phase. The interface is defined by $\phi = 0$. Moreover $\phi_x = \partial\phi/\partial x$ and $\phi_y = \partial\phi/\partial y$ represent partial derivatives with respect to x and y.

The function $F(\phi, \lambda T) = f(\phi) + \lambda T g(\phi)$ is a phenomenological free energy where $f(\phi)$ has the form of a double–well potential, λ controls the coupling between T and ϕ, and the relative height of the free energy minima is determined by T and $g(\phi)$. The function $h(\phi)$ accounts for the release of latent heat. Anisotropy has been introduced in (A.34) by defining $W(\mathbf{n}) = \xi a(\mathbf{n})$ and $\tau(\mathbf{n}) = \tau_o a^2(\mathbf{n})$. τ_o is a time characterizing atomic movement in the solid–liquid interface region, ξ is a length characterizing the width of the interface, and

$$a(\mathbf{n}) = (1 - 3\beta)\left[1 + \frac{4\beta}{1 - 3\beta}\frac{(\phi_x)^4 + (\phi_y)^4}{|\nabla\phi|^4}\right] \tag{A.35}$$

with $a(\mathbf{n}) \in [0, 1]$. a is identical with the function a in (2.8), except that here it is given depending on $\mathbf{n}$ rather than on θ, where $\mathbf{n}$ is the normal vector at the contours of ϕ. For phase-field computations this is convenient, since $\mathbf{n}$ can easily be determined via

$$\mathbf{n} = (\phi_x\hat{\mathbf{x}} + \phi_y\hat{\mathbf{y}})/(\phi_x^2 + \phi_y^2)^{1/2} . \tag{A.36}$$

The constant β parameterizes the deviation of $W(\mathbf{n})$ from ξ and is a measure of the anisotropy strength as in (2.8).

The asymptotic relationships of Karma and Rappel [168, 169] are used to map the phase-field model onto the sharp interface free-boundary problem, where (A.33) and (A.34) reduce to (A.3)–(A.5). In terms of $a(\mathbf{n})$, $\gamma(\mathbf{n}) = \gamma_o a(\mathbf{n})$ and $d(\mathbf{n}) = d_o\left[a(\mathbf{n}) + \frac{\partial^2 a(\mathbf{n})}{\partial\theta^2}\right]$, where θ is the angle between $\mathbf{n}$ and the x-axis, these expressions become (2.8) and $d(\mathbf{n}) = d_o(1 - 15\beta\cos 4\theta)$ in the free-boundary problem (note that $\tan(\theta) = \phi_y/\phi_x$). The parameters of the phase-field model are related to the free-boundary parameters by $\lambda = \xi a_1/d_o$

and $\tau_o = \xi^3 a_1 a_2/(d_o D) + \xi^2 \gamma_0/d_o$. The positive constants a_1 and a_2 depend on the exact form of the phase-field equations. In choosing to simulate particular material characteristics, the experimentally measurable quantities d_o, β, and D are fixed, leaving ξ as a free parameter, which determines λ and τ_o.

Within the simulations reported here fourfold symmetric dendrites are computed in a quarter-infinite space using a finite-element adaptive grid method as in [239, 240]. Solidification is initiated by a small quarter disk of radius R_o centered at the origin. The order parameter is initially set to its equilibrium value $\phi_o(\mathbf{x}) = -\tanh((|\mathbf{x}| - R_o)/\sqrt{2})$ along the interface. The initial temperature is $T = 0$ in the solid. Without loss of generality it is assumed to decay exponentially from $T = 0$ at the interface to $T = -\Delta$ as $\mathbf{x} \to \infty$, where the far-field *undercooling* is given by (A.2). This implies vanishing Δ in (A.6).

Table A.1. Overview over the phase-field models under investigation

Model	$\frac{\partial g(\phi)}{\partial \phi}$	$h(\phi)$	a_1	a_2
1	$1 - \phi^2$	ϕ	$\frac{1}{\sqrt{2}}$	$\frac{5}{6}$
2	$(1 - \phi^2)^2$	ϕ	$\frac{5}{4\sqrt{2}}$	$\frac{47}{75}$
3	$(1 - \phi^2)^3$	ϕ	1.0312	0.52082
4	$(1 - \phi^2)^4$	ϕ	1.1601	0.45448
5	$(1 - \phi^2)^2$	$\frac{15}{8}(\phi - \frac{2}{3}\phi^3 + \frac{1}{5}\phi^5)$	$\frac{5}{4\sqrt{2}}$	0.39809

The different phase-field models studied are summarized in Table A.1. To satisfy the asymptotics, $f(\phi)$ is chosen to be an even function, and $g(\phi)$ and $h(\phi)$ are odd. All of the models employ $f(\phi) = \phi^4/4 - \phi^2/2$. For computational purposes, $g(\phi)$ is chosen such that the two minima of $F(\phi, \lambda T)$ are fixed at $\phi = \pm 1$. Model 1 is a form used by Almgren [11], model 2 by Karma and Rappel [168, 169], and model 5 is the thermodynamically consistent form used by Wang *et al.* [289]. Models 3 and 4 are forms created in particular for this numerical study. Note that model 1 requires λ to be less than $1/\Delta$. Otherwise the $\phi = -1$ state becomes linearly unstable.

Table A.2. Overview of simulation parameters

Δ	L	R_o	Δt	D	d_o	$\tilde{v}$	Pe_{Int}
0.45	1000	17	0.010	3	0.5	0.00545	0.011
0.55	800	15	0.016	2	0.5	0.0170	0.034
0.65	800	15	0.016	1	0.5	0.0469	0.094
0.65	800	15	0.004	2	1.5	0.0469	0.031

In the simulations, the computational domain is an $L \times L$ square box. Computations are performed at $\Delta = 0.65, 0.55$ and 0.45. A summary of the parameters used for each simulation run is given in Table A.2, where $\tilde{v} = v d_o / D$ is the dimensionless tip velocity predicted by linear solvability theory, Δx is the *minimum grid spacing* of the mesh [239, 240], and Δt is the simulation time step. The phase-field parameters are chosen for each model so that they all simulated the same free-boundary problem. For all simulations $\beta = 0.05$, $\xi = 1$ and $\Delta x = 0.39$. As long as $\Delta = 0.45$ or $\Delta = 0.55$ these simulation results reveal that all of the phase-field models studied produce identical results for the entire temporal evolution of the dendrite and also converge to steady state solutions that are within a few percent of those predicted by linear solvability theory.

At $\Delta = 0.65$ (with $d_o = 0.5$), on the other hand, significant quantitative differences between the various phase-field models are encountered. These discrepancies can be attributed to finite corrections of the interfacial Peclet number Pe_{Int} at higher orders of the asymptotic expansion. I.e., one can interpret the deviations as an indication that the solutions do not converge as function of the expansion parameter. Moreover it implies that the phase-field equations do not operate within the sharp interface limit. The universal behavior of the different models can be recovered via a decrease of Pe_{Int}. The price to pay is a reduction of computational efficiency, since the only parameter, which can be controlled within Pe_{Int} is the interfacial width. This interfacial width, however, always has to be resolved by an appropriate grid spacing. At this point a detailed investigation of how to push those finite Pe_{Int}-corrections to still higher orders by careful choices of $f(\phi)$, $g(\phi)$ and $h(\phi)$, seems to be a chance to render phase-field models computationally more efficient. However, to my knowledge such a study is yet unpublished.

A.3 Selected State-of-the-Art Numerical Approaches

Within this section I will present two examples, which display how enhanced computational techniques can be applied to phase-field simulations. The first makes use of an adaptive mesh refinement algorithm and employs dynamical data structures [240] to compute dendritic growth in two dimensions. It is the basis of the second [155], in which the idea of adaptive mesh refinement is extended to 3D simulations of dendritic growth including hydrodynamic flow in the liquid phase. It employs a semi-implicit approximation projection method (SIAPM) along with parallelization. Certainly within the 3D code the precise algorithm for the grid refinement runs somewhat different from that in the 2D case. However, for the sake of simplicity I will discuss grid refinement only for the 2D example. Afterwards I will continue to give an operational description of the additional features concerned with simulating fluid flow in case of the 3D example. These outlines follow [240] and [155], respectively. Together the discussion of these two examples provides concrete

numbers of which performance can be achieved with respect to diffuse interface simulations today.

A.3.1 2D Adaptive Mesh Refinement Computation

The model employed with respect to the numerics presented here is of the general form of model equations (A.33)–(A.34). These equations are solved using
the Galerkin finite element method on dynamically adapting grids of linear,
isoparametric quadrilateral and triangular elements. The grid is adapted dynamically based on an error estimator that utilizes information from both the
ϕ and T fields. In the broadest sense, the algorithm performs functions that
can be divided into two classes. The first deals with the establishment, main-
tenance and updating of the finite element grids, the second with evolving ϕ
and T on these grids according to (A.33) and (A.34).

Maintaining a grid of finite elements on a data structure known as a
quadtree [92, 229, 255] is the main issue of the first class of functions. The
quadtree is a tree-like structure with branches up to a pre-specified level
creating children elements. These are themselves data structures that contain information analogous to the parent, from which they branch, but one
level down. This quadtree element data structure is depicted in Fig. A.1.
Every entry on the quadtree contains information pertaining to a four-noded
isoparametric quadrilateral finite element. This information includes the following:

1. values of ϕ and T at the four nodes,
2. the nodal coordinates of the element,
3. the level of refinement of the element on the quadtree,
4. the value of the current error estimate,
5. the element number, which contains information about the coordinates
   of the element and its level of refinement,
6. an array mapping the four nodes of the element onto the entries of a
   global solution array,
7. pointers to the nearest neighbors of the element sharing a common edge
   at the same level of grid refinement,
8. a variable that determines whether or not an element contains further
   sub-elements termed *child* elements,
9. pointers to the child elements of an element,
10. a pointer to the *parent* element, from which an element originates.

A parent element and its four child elements are referred to as a *family*.
Refinement produces a finer mesh within the confines of the original parent
grid by bisecting each side. Unrefinement, which consists of fusing the four
child elements back into the parent, has the opposite effect, locally creating a coarser mesh. Both refinement and unrefinement proceed via dynamic
memory allocation, making the code scalable. Unrefinement can occur only
if the child elements do not possess further children of their own. Also, in

order to avoid regions of different refinement bordering each other, the restriction that any two neighboring quadrilateral elements may be separated by no more than one level of refinement is imposed. Moreover one defines a level of refinement of an element l_e, so that a uniform grid at refinement level l_e would contain $2^{l_e} \times 2^{l_e}$ grid points in a physical domain $L_B \times L_B$.

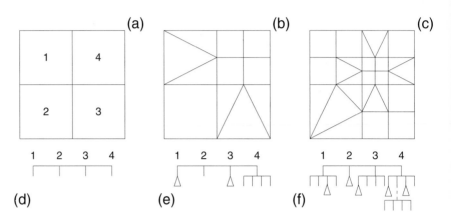

Fig. A.1. An illustration of the quadtree element data structure. (a) An element and four child elements. (b) Splitting of the children. (c) Splitting of the childrens' children. (d)–(e) Evolution of branching in the quadtree. Branches with triangles indicate square elements which are bridged with triangular or rectangular elements. Figure following [240].

Cases, in which an element has no children, a missing neighbor, or no parent are handled by null pointers. The latter case occurs only for the root of the quadtree. All elements at a given level of refinement on the quadtree are "strung" together by a linked list of pointers, referred to as the *G-list*. There are as many *G*-lists as there are levels of refinement in the quadtree. Each pointer in the *G*-list accesses the location in memory assigned to one element of the quadtree. The purpose of the *G*-list is to allow traversing of the quadrilateral elements sequentially rather than recursively. The latter procedure would be memory intensive and relatively slow.

In addition the code maintains two independent grids representing special linear isoparametric triangular and rectangular elements. These elements are used to connect the extra nodes, which arise when two or more quadrilateral elements of different refinement levels border each other. These element types are referred to as *bridging elements*. They are maintained as two linked lists of derived data types, one containing information about triangular elements, the other rectangular. Elements of both grids include the following information:

1. the values of ϕ and T at the three nodes (four for rectangles) of the element,

2. the nodal coordinates,
3. node numbers that map the nodes of the element onto the global solution
   array.

The main set of operations performed on the grids described above concern refinement of the finite element mesh as a whole. The refinement process
is performed only on the quadrilateral mesh. The triangular and rectangular
grids are established after this process is completed. To refine the grid the
code traverses the elements of the quadtree, refining (unrefining) any element,
for which the error estimate is above (below) a critical value $\sigma_h(\sigma_l)$. Fusion
of four quadrilateral elements can only occur, if all four error estimates are
below the critical value σ_l, where $\sigma_l < \sigma_h$. Simulations show [240] that for
$\sigma_l = \sigma_h$ the grid starts to oscillate, i.e. a large numbers of elements become
alternatively refined at one time step, then unrefined at the next.

The processes described so far are grouped into modules that encapsulate
various related tasks. They can cross-reference the data and instructions of
each other. The module highest in hierarchy contains the definition of the
quadtree data structure as well as routines that construct the initial uniform
grid, refine and unrefine individual quadrilateral elements and impose the initial conditions. Another module constructs the G-lists. It contains routines
that construct the initial G-list from an initially uniform quadtree data structure. Moreover it adds (deletes) element pointers from the G-list as elements
are created (deleted) from the quadtree. Another module accessing the data
structure of the previous ones has the role of creating the triangular and
rectangular element grids. It contains definitions for creating triangular and
rectangular elements data structures and routines that search the quadtree,
building the linked lists of triangles and rectangles that make up these grids.

The second class of functions is concerned with updating of the finite
element grids. With respect to this point the final modul runs the integration
of (A.33) and (A.34). This module performs four main processes:

1. It maps the internal element node numbers to the indices of a global
   solution vector.
2. It advances the T and ϕ field vectors by N_r time steps on the finite
   element grids defined above.
3. It calculates an error estimate for each element of the quadtree, based on
   an error estimate of the quadrilateral elements.
4. It invokes the routines to refine the grid according to the error estimator.

Steps (1)-(4) are repeated until a sufficient time evolution of the microstructure is established. The variable N_r is set, so that the interface remains within the regions of fine meshing between regriddings. The latter is
typically done every 100 time steps. Step (1) involves searching for all elements and the respective neighbors, as well as assigning each node a unique
number, if it has a counterpart on a global solution vector.

The finite element discretization of (A.33) and (A.34) is done using Galerkin's weighted residual method [78]. The method begins by assuming that ϕ and T are interpolated within an element as

$$\phi^e = \sum_{i=1}^{N} \phi_i^e N_i(x,y) \qquad T^e = \sum_{i=1}^{N} T_i^e N_i(x,y) \qquad \text{(A.37)}$$

where ϕ_i^e and T_i^e are the field values at the N nodes of the element e and their interpolated values in its interior. The functions $N_i(x,y)$ are standard linear interpolation functions appropriate for the used element [303]. They satisfy

$$N_i(x_j, y_j) = \delta_{i,j} , \qquad \text{(A.38)}$$

where $\delta_{i,j}$ is the Kronecker delta. Rewriting the differential equations for ϕ in (A.33) and (A.34) as $L_\phi \phi = 0$, as well as of the T-equation as $L_T T = 0$, the Galerkin method requires that

$$\int_{\Omega_e} N_i(x,y) L_\phi \phi^e(x,y) \mathrm{d}x\mathrm{d}y = 0 \qquad \text{(A.39)}$$

$$\int_{\Omega_e} N_i(x,y) L_T T^e(x,y) \mathrm{d}x\mathrm{d}y = 0 ,$$

for $i = 1,2,3,\ldots,N$, where Ω_e represents the area of an element e. Substituting (A.37) into (A.39), one obtains two linear algebraic equations for ϕ_i and T_i, $i = 1,2,3,\ldots,N$ in the element e.

The next step is to define $\Phi^e = (\phi_1, \phi_2, \phi_3, \cdots, \phi_N)^{\mathrm{T}}$ and $\mathbf{Y}^e = (T_1, T_2, T_3, \cdots, T_N)^{\mathrm{T}}$, where the superscript T denotes the transpose, making Φ^e and $\mathbf{T}^e$ column vectors. Then the linear algebraic statement of the finite element form of (A.33) and (A.34) reads

$$\hat{\mathbf{C}}(\phi^e)\frac{\mathrm{d}\phi^e}{\mathrm{d}t} = \left(\hat{\mathbf{A}} + \mathbf{E}\right)\phi_n^e + \mathbf{F}^e(\lambda) \qquad \text{(A.40)}$$

$$\mathbf{C}\frac{\mathrm{d}\mathbf{T}^e}{\mathrm{d}t} = D\mathbf{A}\mathbf{T}^e + \frac{1}{2}\mathbf{C}\frac{\mathrm{d}\phi^e}{\mathrm{d}t} ,$$

where the matrices $\mathbf{C}$, $\hat{\mathbf{C}}$, $\mathbf{A}$, $\hat{\mathbf{A}}$ and $\mathbf{E}$ and the vector $\mathbf{F}^e(\lambda)$ are given by

$$\mathbf{C} = \int_{\Omega_e} \mathbf{N}^{\mathrm{T}}\mathbf{N}\mathrm{d}x\mathrm{d}y , \qquad \text{(A.41)}$$

$$\hat{\mathbf{C}} = \int_{\Omega_e} \mathbf{N}^{\mathrm{T}}\mathbf{N} A^2(\theta(\phi^e))\mathrm{d}x\mathrm{d}y , \qquad \text{(A.42)}$$

$$\mathbf{A} = -\int_{\Omega_e} \left(\mathbf{N}^{\mathrm{T}}\mathbf{N_x} + \mathbf{N}^{\mathrm{T}}\mathbf{N_y}\right)\mathrm{d}x\mathrm{d}y , \qquad \text{(A.43)}$$

$$\hat{\mathbf{A}} = -\int_{\Omega_e} \left(\mathbf{N}^{\mathrm{T}}\mathbf{N_x} + \mathbf{N}^{\mathrm{T}}\mathbf{N_y}\right) A^2(\theta(\phi^e))\mathrm{d}x\mathrm{d}y , \qquad \text{(A.44)}$$

$$\mathbf{E} = -\int_{\Omega_e} \left(\mathbf{N}^\mathsf{T} \mathbf{N_x} - \mathbf{N}^\mathsf{T} \mathbf{N_y} \right) A(\theta(\phi^e)) \omega(\theta(\phi^e)) \mathrm{d}x \mathrm{d}y \;, \tag{A.45}$$

$$\mathbf{F}^e(\lambda) = \int_{\Omega_e} \mathbf{N}^\mathsf{T} f(\phi^e, U^e; \lambda) \mathrm{d}x \mathrm{d}y \;. \tag{A.46}$$

Here $\mathbf{N_x}$, $\mathbf{N_y}$ denote the partial derivatives of the vector of shape functions with respect to x and y, respectively. The function $A(\theta(\phi^e))$ is given by (A.35). $\omega(\theta)$ is proportional to the derivative of $A(\theta)$. It reads:

$$\omega(\theta(\phi^e)) = 16\beta \frac{\tan\theta(1 - \tan^2\theta)}{(1 + \tan^2\theta)^2} \;. \tag{A.47}$$

A lumped formulation for the matrices $\mathbf{C}$ and $\hat{\mathbf{C}}$ [78] is employed. In this procedure, the row vector of shape functions $\mathbf{N}$ in (A.41) is replaced by the identity row vector $\mathbf{I} = (1, 1, 1, \cdots)$. The resulting matrix $\mathbf{C}$ consists of identical columns, each of which contains the element $N_i(x, y)$ at the position of the ith row. A lumped term is defined as a diagonal matrix of entries which take the values

$$L_c = \frac{1}{\text{nodes}} \sum_{i=1}^{\text{nodes}} \int_{\Omega_e} N_i(x, y) \mathrm{d}x \mathrm{d}y \;. \tag{A.48}$$

The use of a lumped matrix for $\mathbf{C}$ enables the assembly of a diagonal matrix for the left hand side of (A.40), stored as a one-dimensional vector. (Two-dimensional arrays would be required if the consistent formulation for the assembly of the $\mathbf{C}$ matrices were used. However, microstructures evolving at low undercooling can produce interfaces with over $2 \cdot 10^5$ elements, making the storing of $2 \cdot 10^5 \times 2 \cdot 10^5$ matrices impossible.)

The global ϕ (obtained after assembly of the element equations in (A.40)) is time-stepped using a forward difference (explicit) time scheme. For each time step of the ϕ field, the global T field is solved iteratively using a Crank–Nicholson scheme. Convergence of T is obtained within a few iterations.

Regridding is based on an error estimator function discussed next. It is obtained following Zienkiewicz and Zhu [303], based on the differences between calculated and smoothed gradients of the ϕ and T fields. In particular a *composite field*

$$\Psi = \phi + \gamma T \tag{A.49}$$

is defined, for which γ is a constant. This definition enables a regridding in accordance to the requirements of both the ϕ and T field, as opposed to using merely the gradients of the ϕ-field in establishing the grid [39]. Since ϕ and T have to be determined rather than their gradients, one does not expect the gradient of Ψ to be continuous across element boundaries, due to the order of the interpolation used. Thus the difference between the calculated and the smoothed gradients, which are both continuous across element boundaries, can be expected to provide a reasonable estimate of error. This method appropriately meshes regions of both steep gradients and regions where the ϕ and T fields change rapidly.

The error estimator function $\mathbf{e}$ is defined as

$$\mathbf{e} = \mathbf{q}_s - \mathbf{q}_c \,, \tag{A.50}$$

where $\mathbf{q}_c$ and $\mathbf{q}_s$ are the calculated and smoothed gradients of Ψ, respectively. To determine $\mathbf{q}_s$, one assumes it to be interpolated in the same way as the ϕ and T fields, namely

$$\mathbf{q}_s = \mathbf{N}\mathbf{Q}^s \,, \tag{A.51}$$

where $\mathbf{N}$ is the row vector of element shape functions and $\mathbf{Q}^s$ a 4×2 matrix, which columns represent the nodal values of fluxes of Ψ in the x and y direction, respectively. To find $\mathbf{Q}^s$, Galerkin's method is employed, minimizing the weighted residual

$$\int_{\Omega_e} \mathbf{N}^T \mathbf{e} \, d\Omega_e = \int_{\Omega_e} \mathbf{N}^T (\mathbf{N}\mathbf{Q}^s - \mathbf{q}_c) d\Omega = 0 \,. \tag{A.52}$$

This calculation is simplified by lumping the left hand side of (A.52), leading to

$$\left(\int_{\Omega_e} \mathbf{N}^T \mathbf{Id}\Omega \right) \mathbf{Q}^s = \int_{\Omega_e} \mathbf{N}^T \mathbf{q}_c d\Omega \,. \tag{A.53}$$

Assembling (A.53) for all quadrilateral elements yields an equation for the smoothed gradients $\mathbf{Q}^g$ of the global field Ψ, at all element nodes, of the form

$$\mathbf{D}\mathbf{Q}^g = b \,, \tag{A.54}$$

where $\mathbf{D}$ is a diagonal matrix, due to "masslumping", and $\mathbf{Q}^g$ is a $N \times 2$ matrix for the global, smoothed flux.

For the actual error updating on the elements of the quadtree one employs the normalized error

$$E_e^2 = \frac{\int_{\Omega_e} |(\mathbf{q}_s - \mathbf{q}_c)|^2 d\Omega_e}{\sum_e \int_{\Omega} |\mathbf{q}_s|^2 d\Omega} \,. \tag{A.55}$$

The domain of integration Ω in the denominator denotes the entire domain of the problem. Thus E_e^2 gives the contribution of the local element error relative to the total smoothed error calculated over the entire grid.

The CPU time of this approach scales with the arc length of the problem under investigation. For most computations this implies an increase in computational efficiency compared to computations on a uniform grid, which scale with the size of the required grid itself. E.g., for $\Delta = 0.70$, $D = 2$, $dt = 0.016$, λ chosen to simulate $\beta = 0$ and system size 800×800 with $\Delta x_{min} = 0.4$, simulations of 10^5 time steps take approximately 15 CPU-hours on a Sun UltraSPARC 2200 workstation with 512 MHz.

A.3.2 3D Simulations Including Fluid Flow

The phase-field model underlying the computational approach presented within this section employs the same phase-field model equations as the previous section. To extend it to hydrodynamic flow in the liquid phase a method proposed by Beckermann *et al.* [26] is employed. Within this approach a phase average Φ_k of a variable Φ for phase k over volume ΔV is defined as

$$\Phi_k = \frac{2 \int_{\Delta V} X_k \Phi dV}{1 - \phi} , \qquad (A.56)$$

where $X_k \in \{0, 1\}$ is an existence function. The formulation ensures that the fluid velocity is extinguished in the solid, and further that the shear stress at the liquid–solid interface is handled correctly.

The governing equations for simulating dendritic growth with fluid flow are the **mixture continuity equation**:

$$\nabla \cdot \left[\frac{1 - \phi}{2} \mathbf{u} \right] = 0 , \qquad (A.57)$$

where $\mathbf{u}$ is the velocity vector, the **averaged momentum equation**:

$$\frac{\partial}{\partial t} \left[\left(\frac{1 - \phi}{2} \right) \mathbf{u} \right] + \mathbf{u} \cdot \nabla \left[\left(\frac{1 - \phi}{2} \right) \mathbf{u} \right] + \left(\frac{1 - \phi}{2} \right) \nabla p$$
$$= \nu \nabla^2 \left[\left(\frac{1 - \phi}{2} \right) \mathbf{u} \right] - \nu \frac{h^2 (1 - \phi^2)(1 + \phi)}{8\delta^2} \mathbf{u} , \qquad (A.58)$$

where t is time, p the pressure, ν the kinematic viscosity, $\delta = W_0/\sqrt{2}$ the characteristic interface width, and h a constant ($=2.757$), which ensures that the interface shear stress is correct for a simple shear flow (see Beckermann *et al.* [26]), and **the averaged energy conservation equation** in terms of the dimensionless temperature T:

$$\frac{\partial T}{\partial t} + \left(\frac{1 - \phi}{2} \right) \mathbf{u} \cdot \nabla T = D \nabla^2 T + \frac{1}{2} \frac{\partial \phi}{\partial t} . \qquad (A.59)$$

The phase averages of velocity and pressure are used for deriving the mixture continuity equation, averaged liquid momentum equation and averaged energy conservation equation. These are solved together with the **3D phase-field evolution equation**, which is a straightforward extension of (A.34).

Here the 3D flow equations are solved using the SIAPM by Gresho [136]. SIAPM is a predictor-corrector method, which can solve (A.57) and (A.58) effectively, especially for large 3D problems, because it uses relatively small amounts of memory. The velocity degrees of freedom are solved in a segregated form. Pressure is updated using a projection method. For a detailed discussion of the algorithm, the reader is referred to the original paper [136]. The algorithm consists of three steps, which are summarized briefly in the following:

1. Compute an intermediate velocity $\tilde{\mathbf{u}}^{n+1}$ from

$$\left(\frac{1}{\Delta t}\mathbf{M} - \frac{1}{2}\mathbf{K} + \mathbf{F}\right)\tilde{u}_i^{n+1} = \left(\frac{1}{\Delta t}\mathbf{M} + \frac{1}{2}\mathbf{K}\right)u_i^n$$
$$-\mathbf{A}(\mathbf{u}^n)u_i^n - \mathbf{G}_i\mathbf{p}^n \ , \tag{A.60}$$

where $\tilde{\mathbf{u}}_i^{n+1}$ is the vector of nodal values of the intermediate velocity component i at time step $n + 1$, $\mathbf{u}_i^n$ is the corresponding vector at time step n, and $\mathbf{p}^n$ is the vector of nodal pressures at time step n. The coefficient matrices are defined in terms of the velocity shape functions $\mathbf{N}$ as follows:

$$\mathbf{M} = \int_\Omega \frac{(1-\phi)}{2}\mathbf{N}^\mathsf{T}\mathbf{N}d\Omega \tag{A.61}$$

$$\mathbf{K} = \int_\Omega \nu\frac{(\phi-1)}{2}\left(\frac{\partial \mathbf{N}^\mathsf{T}}{\partial \mathbf{x}}\frac{\partial \mathbf{N}}{\partial \mathbf{x}} + \mathbf{I}\frac{\partial \mathbf{N}^\mathsf{T}}{\partial x_k}\frac{\partial \mathbf{N}}{\partial x_k}\right)d\Omega \tag{A.62}$$

$$\mathbf{F} = \int_\Omega \frac{\nu h(1-\phi^2)(1+\phi)}{8\delta^2}\mathbf{N}^\mathsf{T}\mathbf{N}d\Omega \tag{A.63}$$

$$\mathbf{A}(\mathbf{u}^n) = \int_\Omega \frac{(1-\phi)}{2}\mathbf{N}^\mathsf{T}u_k^n\frac{\partial \mathbf{N}}{\partial x_k}d\Omega \tag{A.64}$$

$$\mathbf{G}_i = \int_\Omega \frac{(1-\phi)}{2}\mathbf{N}^\mathsf{T}\frac{\partial \mathbf{N}}{\partial x_i}d\Omega \ . \tag{A.65}$$

2. The velocity field found in the first step is generally not divergence-free. The next step corrects the pressure to obtain an approximately divergence-free velocity field by solving a Poisson equation for $\Delta\mathbf{p}^{n+1} = \mathbf{p}^{n+1} - \mathbf{p}^n$:

$$\mathbf{L}\Delta\mathbf{p}^{n+1} = -\frac{1}{\Delta t}\mathbf{D}\left(\tilde{\mathbf{u}}^{n+1} - \mathbf{u}^n\right) \ , \tag{A.66}$$

where

$$\mathbf{L} = \int_\Omega \frac{(1-\phi)}{2}\mathbf{I}\frac{\partial \mathbf{N}^\mathsf{T}}{\partial x_k}\frac{\partial \mathbf{N}}{\partial x_k}d\Omega \tag{A.67}$$

$$\mathbf{D} = \int_\Omega \frac{(1-\phi)}{2}\mathbf{N}^\mathsf{T}\frac{\partial \mathbf{N}}{\partial \mathbf{x}}d\Omega \ . \tag{A.68}$$

$\mathbf{p}^{n+1}$ is then updated according to

$$\mathbf{p}^{n+1} = \mathbf{p}^n + \Delta\mathbf{p}^{n+1} \ . \tag{A.69}$$

3. Finally, the projected velocity $\mathbf{u}^{n+1}$ is computed in a corrector step by solving

$$\mathbf{u}^{n+1} = \tilde{\mathbf{u}}^{n+1} - \Delta t\mathbf{M_L}^{-1}\mathbf{G}\Delta\mathbf{p}^{n+1} \ , \tag{A.70}$$

where

$$\mathbf{M_L} = \int_\Omega \frac{1-\phi}{2}\mathbf{N}d\Omega \ . \tag{A.71}$$

The computations are started using an initial velocity field $\mathbf{u}_0$ determined from the boundary conditions. In order to obtain a field $\mathbf{u}_0$, whose discrete divergence is close to zero, the pressure is adapted by iteration. The initial pressure is zero everywhere.

Equations (A.60) and (A.66) are solved by the conjugate gradient (CG) method with diagonal preconditioning [217]. The SIAPM can calculate the velocity field for large 3D problems much faster than fully implicit time-stepping methods, because convergence for (A.60) is reached within a few iterations. Moreover, the number of degrees of freedom of (A.66) is one. The CG iteration for (A.66) converges more slowly, typically within 50-200 iterations.

The averaged energy equation (A.59) is solved using the CG method with diagonal preconditioning as well. Streamline upwind schemes [45] are employed for the convection terms in (A.60) and (A.59). The 3D phase-field equation is a nonlinear system. In order to solve the system implicitly, an iterative method such as the Newton–Raphson method is required. Alternatively an explicit time-stepping scheme can be employed. Stable solutions are obtained from the explicit scheme, since the variation of the ϕ field exists only in the interface region and a sufficiently small time increment Δt can be used.

In analogy to the previous section the algorithm is implemented along with adaptive grid refinement. Moreover it is parallelized on the basis of the Charm++FEM framework [158].

Based on this implementation in [155] for a set of problem parameters corresponding to SCN, i.e. $\Delta t = 0.016$, $W = 1$, $\tau_0 = 1$, $\lambda = 6.383$, $d_0(= a_1 \frac{W}{\lambda}) = 0.139$, $\nu = 92.4$ and $Pr = 23.1$ computations of 100 time steps of a growing 3D dendrite are reported to require 190 hours of CPU for an underlying minimum grid spacing of $\Delta x_{\min} = 0.4$. 80% to 90% of the CPU time were spent for the update of the velocity and the pressure field [155].

To my knowledge these 3D simulations of SCN including fluid flow are the most extensive computations of dendritic growth so far. Everything beyond still remains an open challenge – from point of view of numerics as well as from point of view of modeling, which might help to simplify numerics.

References

1. T. Abel, E. Brener, H. Müller-Krumbhaar: Phys. Rev. E **55**, 7789 (1997)
2. T. Abel: *Phasenfeldmodelle für Wachstumsprozesse in Multikomponenten- und Multiphasensystemen.* PhD Thesis, Research Center Jülich, Jülich (1998)
3. *Handbook of Mathematical Funktions* ed. by M. Abramowitz, I.A. Stegun, 10th ed. (Wiley, New York 1972)
4. G. Adam, O. Hittmair: *Wärmetheorie*, 3rd ed. (Vieweg, Braunschweig 1988)
5. S. Akamatsu, G. Favier, T. Ihle: Phys. Rev. E **51**, 4751 (1995)
6. M. Albrecht, S. Christiansen, J. Michler, W. Dorsch, H.P. Strunk, P.O. Hansson, E. Bauser: Appl. Phys. Lett. **67**, 1232 (1995)
7. S.M. Allen, J.W. Cahn: Acta Metall. **27**, 1085 (1979)
8. F.J. Almgren, J.E. Taylor, L. Wand: SIAM J. Cont. Opt. **31**, 386 (1993)
9. R. Almgren, W. S. Dai, V. Hakim: Phys. Rev. Lett. **71**, 3461 (1993)
10. R. Almgren: J. Comp. Phys. **106**, 337 (1993)
11. R. Almgren: SIAM J. Appl. Math. **59**, 2086 (1999)
12. D.M. Anderson, G.B. McFadden, A.A. Wheeler: Phys. Fluids **9**, 1870 (1997)
13. D.M. Anderson, G.B. McFadden, A.A. Wheeler: Ann. Rev. Fluid Mech. **30**, 139 (1998)
14. D.M. Anderson, G.B. McFadden, A.A. Wheeler: NISTIR 6237 (1998)
15. D.M. Anderson, G.B. McFadden, A.A. Wheeler: Physica D **135**, 175 (2000)
16. D.M. Anderson, G.B. McFadden, A.A. Wheeler: Physica D **144**, 154 (2000)
17. D.M. Anderson, G.B. McFadden, A.A. Wheeler: In *Interfaces for the Twenty-First Century*, ed. by M.K. Smith, M.J. Miksis, G.B. McFadden, G.P. Neitzel, D.R. Canright (Imperial College Press, London 2002), p. 131
18. D.M. Anderson, G.B. McFadden, A.A. Wheeler: Physica D **151**, 305 (2001)
19. R. Aris: *Vector, Tensors, and the Basic Equations of Fluid Dynamics* (Dover, New York 1962), p. 84
20. G.S. Bales, A. Zangwill: Phys. Rev. B **41**, 5500 (1990)
21. A.-L. Barabasi, H. E. Stanley: *Fractal Concepts in Surface Growth*, 1st ed. (Cambridge University Press, Cambridge 1995)
22. M.Barber, A.Barbieri, J.S. Langer: Phys. Rev. A **36**, 3340 (1987)
23. M.N. Barber, D. Singleton: J. Math. Soc. B **36**, 325 (1994)
24. W. Blanke (ed.): *Thermophysikalische Stoffgrößen*, 1st ed. (Springer-Verlag, Berlin 1989), p. 155
25. J. Bechhoefer: *Directional Solidification at the Nematic–isotropic Interface.* PhD Thesis, The University of Chicago, Chicago (1988)
26. C. Beckermann, H.-J. Diepers, I. Steinbach, A. Karma, X. Tong: J. Comp. Phys. **154**, 468 (1999)
27. M. Ben Amar, B. Moussalam: Phys. Rev. Lett. **60**, 317 (1988)
28. M. Ben Amar, P. Pelce: Phys. Rev. A **39**, 4263 (1989)
29. E. Ben-Jacob, N. Goldenfeld, J.S. Langer, G. Schön: Phys. Rev. A **29**, 330 (1984)

30. E. Ben-Jacob, N. Goldenfeld, J.S. Langer, G. Schön: Phys. Rev. Lett. **51**, 1930 (1983)
31. C.M. Bender, S.A. Orszag: *Advanced Mathematical Methods for Scientists and Engineers* (McGraw-Hill, New York 1978)
32. K.W. Benz: private communication (Freiburg 2002)
33. J.J. Binney: *The Theory of Critical Phenomena: An Introduction to the Renormalization Group*, 1st repr. with corr. (Clarendon, Oxford 1999)
34. P. Bouissou, B. Perrin, P. Tabeling: Phys. Rev. A **40**, 509 (1989)
35. W.J. Boettinger, J.A. Warren: Met. Trans. A **27**, 657 (1996)
36. W.J. Boettinger, J.A. Warren: J. Cryst. Growth **200**, 583 (1999)
37. A. Bösch: *Phasenfeld-Modelle für Mehrkomponentensysteme*. Diploma Thesis, Research Center Jülich, Jülich (1993)
38. A. Bösch, H. Müller–Krumbhaar, O. Shochet: Z. Phys. **97**, 367 (1995)
39. R.J. Braun, B.T. Murray: J. Cryst. Growth **174**, 41 (1997)
40. R.J. Braun, J.W. Cahn, G.B. McFadden, H.E. Rushmeier, A.A. Wheeler: Acta Mater. **46**, 1 (1998)
41. A. J. Bray: Adv. Phys. **32**, 357 (1994)
42. E. Brener, V. I. Melnikov: Adv. Phys. **40**, 53 (1991)
43. E. Brener: Phys. Rev. Lett. **71**, 3653 (1993)
44. E. Brener, D. Temkin: Phys. Rev. E **51**, 351 (1995)
45. A.N. Brooks, T.J.R. Hughes: Comp. Methods Appl. Mech. Eng. **32**, 199 (1982)
46. R. Brower, D. Kessler, J. Koplik, H. Levine: Phys. Rev. Lett. **51**, 1111 (1983)
47. W.K. Burton, N. Cabrera, F.C. Frank: Phil. Trans. Roy. Soc. **243**, 299 (1951)
48. G. Caginalp, P. Fife: Phys. Rev. B **33**, 7792 (1986)
49. G. Caginalp: Arch. Rat. Mech. Anal. **92**, 205 (1986)
50. G. Caginalp: SIAM J. Appl. Math. **48**, 506 (1988)
51. C. Caginalp, J. Jones: App. Math. Lett. **4**, 79 (1991)
52. G. Caginalp, X. Chen: In *On The Evolution Of Phase Boundaries*, ed. by M.E. Gurtin and G.B. McFadden (Springer, New York, 1992), p. 1
53. G. Caginalp, W. Xie: Phys. Rev. E **48**, 1897 (1993)
54. J.W. Cahn, J. E. Hilliard: J. Chem. Phys. **28**, 258 (1958)
55. J.W. Cahn, J.E. Hilliard: J. Chem Phys. **31**, 688 (1959)
56. J.W. Cahn: Acta Metall. **8**, 554 (1960)
57. J.W. Cahn: Acta Metall. **9**, 795 (1961)
58. J.W. Cahn, D.W. Hoffmann: Acta Met. **22**, 1205 (1974)
59. J.W. Cahn, W.C. Carter: Met. Trans. **27A**, 1431 (1996)
60. J.W. Cahn, P. Fife, O. Penrose: Acta Mater. **45**, 4397 (1997)
61. R.W. Cahn, P. Haase: *Physical Metallurgy*, 4th ed. (North Holland, Amsterdam 1996)
62. R.W. Cahn, P. Haase, E.J. Kramer: *Materials and Technology*, 1st ed. (VCH, Weinheim 1991)
63. D. Caraballo: *A Variational Scheme for the Evolution of Polycrystals by Curvature*. PhD Thesis, Department of Mathematics, Princeton University (Princeton, 1997)
64. B. Caroli, C. Caroli, B. Roulet: In *Solids Far From Equilibrium*, ed. by G. Godrèche (Cambridge University Press, Cambridge 1992), p. 155
65. M. Carrard, M. Gremaud, M. Zimmermann, W. Kurz: Acta Metall. **40**, 983 (1992)
66. Y.C. Chang, T.Y. Hou, B. Merriman, S. Osher: J. Comput. Phys. **124**, 449 (1996)
67. Ch. Charach, P.C. Fife: SIAM J. Appl. Math. **58**, 1826 (1998)
68. Ch. Charach, P.C. Fife: Open Syst. Inf. Dyn. **5**, 99 (1998)
69. Ch. Charach, C.K. Chen, P.C. Fife: J. Stat. Phys. **95**, 1141 (1999)

70. S. Chen, B. Merriman, S. Osher, P. Smereka: J. Comput. Phys. **135**, 8 (1997)
71. S. Chen, B. Merriman, M. Kang, R.E. Caflisch, C. Ratsch, L.-T. Cheng, M. Gyure, R.P. Fedkiw, C. Anderson, S. Osher: J. Comput. Phys. **167**, 475 (2001)
72. D.L. Chopp: J. Comput. Phys. **162**, 104 (2000)
73. S. Clarke, D.D. Vvendensky: Phys. Rev. Lett. **58**, 2235 (1987)
74. J.B. Collins, H. Levine: Phys. Rev. B **31**, 6119 (1985)
75. M. Conti: Phys. Rev. E **56**, 3197 (1997)
76. M. Conti: Phys. Rev. E **55**, 701 (1997)
77. M. Conti: Phys. Rev. E **55**, 765 (1997)
78. R. Cook, D. Malkus, M. Plesha: *Concepts and Applications of Finite Element Analysis*, 3rd ed. (Wiley, New York 1989)
79. J. B. Collins, H. Levine: Phys. Rev. B **31**, 6119 (1985)
80. S.R. Coriell, D. Turnbull: Acta Metall. **30**, 2135 (1982)
81. S.R. Coriell, R.S. Sekerka: J. Cryst. Growth **61**, 499 (1983)
82. S.R. Coriell, R.F. Boisvert, G.B. McFadden, L.N. Brush, J.J. Favier: J. Cryst. Growth **140**, 139 (1994)
83. S. R. Coriell, G. B. McFadden, R. F. Sekerka and W. J. Boettinger: J. Cryst. Growth **191**, 573 (1998)
84. D.P. Corrigan, M.B. Koss, J.C. LaCombe, K.D. de Jager, L.A. Tennenhouse, M.E. Glicksman: Phys. Rev. E **60**, 7217 (1999)
85. M.E. Curtin, D. Polignone, J. Vinals: Math. Models. Methods Appl. Sci. **6**, 815 (1996)
86. A.G. Cullis, D.J. Robbins, A.J. Pidduck, P.W. Smith: Mater. Res. Soc. Symp. Proc. **280**, 383 (1993)
87. A.G. Cullis: Mater. Res. Soc. Symp. Proc. **399**, 303 (1996)
88. J.A. Dantzig, L.S. Chao: In *Proc. 10th U.S. Nat. Cong. of Appl. Mech.*, ed. by J. Lamb (1986), p. 249
89. S.H. Davis: *Theory of Solidification*, 1st ed. (Cambridge University Press, Cambridge 2001)
90. S. de Cheveigne, C. Guthmann, M.M. Lebrun: J. Phys. Paris **47**, 2095 (1986)
91. S.R. de Groot, P. Mazur: *Non-Equilibrium Thermodynamics*, 1st ed. (Dover, New York 1984)
92. P.R.B. Devloo: *An H-P Adaptive Finite Element Method for Steady Compressible Flow*. Ph.D Thesis, The University of Texas at Austin, Austin, TX (1987)
93. C. Domb, M.S. Green: *Phase Transitions and Critical Phenomena*, Vol. **5** (Academic Press, London 1976)
94. W. Dorsch, S. Christiansen, M. Albrecht, P.O. Hansson, E. Bauser, H.P. Strunk: Surf. Sci. **896**, 331 (1994)
95. W. Dorsch, B. Steiner, M. Albrecht, H.P. Strunk, H. Wawra, G. Wagner: J. Cryst. Growth **183**, 305 (1998)
96. F. Drolet, K. R. Elder, Martin Grant, J. M. Kosterlitz: Phys. Rev. E **61**, 6705 (2000)
97. D.J. Eaglesham, M. Cerullo: Phys. Rev. Lett. **64**, 1943 (1990)
98. Ch. Eck, P. Knabner, S. Korotov: 'A two-scale model for the computation of solid-liquid phase transitions with dendritic microstructure'. Preprint No. 278, Institute for Applied Mathematics, University Erlangen-Nürnberg (2001)
99. W. Eckhaus: *Matched Asymptotic Expansions and Singular Perturbations* (North Holland, Amsterdam 1973)
100. K R. Elder, F. Drolet, J.M. Kosterlitz, M. Grant: Phys. Rev. Lett. **72**, 677 (1994)
101. K.R. Elder, J.D. Gunton, M. Grant: Phys. Rev. E **54**, 6476 (1996)

102. K.R. Elder, M. Grant, N. Provatas, J.M. Kosterlitz: Phys. Rev. E **64**, 021604 (2001)
103. The appendix of this chapter follows closely the appendices of Elder et al. [102].
104. G. Ehrlich: Surface Sci. **63**, 422 (1977)
105. H.F. El–Nashar, H.A. Cerdeira: Physica A **283**, 6 (2000)
106. H. Emmerich, K. Kassner, C. Misbah, T. Ihle: J. Phys.: Condens. Matter **11**, 9985 (1999)
107. H. Emmerich, D. Schleussner, K. Kassner, T. Ihle: J. Phys.: Condens. Matter **11**, 8981 (1999)
108. H. Emmerich, K. Kassner, T. Ihle, A. Weiß: J. Cryst. Growth **211**, 43 (2000)
109. H. Emmerich: Phys. Rev. B **65**, 233406 (2002)
110. H. Emmerich: accepted for publication in Continuum Mech. Therm.
111. H. Emmerich, M. Jurgk, R. da Silva de Siquieri: in preparation
112. M.A. Eshelman, V. Seetharaman, R. Trivedi: Acta Metall. **36**, 1165 (1988)
113. M. Fabbri, V.R. Voller: J. Comp. Phys. **130**, 256 (1997)
114. R. Fedkiw, T. Aslam, B. Merriman, S. Osher: J. Comput. Phys. **152**, 457 (1999)
115. P.C. Fife: In *CBMS-NSF Regional Conference Series in Applied Mathematics* **53** (SIAM, Philadelphia 1988), p. 11
116. P.C. Fife, O. Penrose: Electron. J. Diff. Equations **1**, 1 (1995)
117. A detailed investigation of the 'classical' problem of density change flow without generalized boundary conditions was carried out by T. Fischalek, Otto-von-Guericke University Magdeburg (2000). Results presented at the TMR network 'Pattern, Noise and Chaos' meeting Leiden, July 2000. The calculations presented in this section are based on this work.
118. M. Fisher: Rev. Mod. Phys. **46**, 597 (1974)
119. G. Fix: Research Notes in Mathematics **2**, (1983)
120. G.J. Fix: In *Free Boundary Problems: Theory and Applications, Vol. II*, ed. by A. Fasano and M. Primicerio (Pitman, Boston 1983), p. 580
121. E. Fried, M.E. Gurtin: Physica D **72**, 287 (1994)
122. E. Fried, M.E. Gurtin: Physica D **91**, 143 (1996)
123. H. Gao: J. Mech. Phys. Solids **42**, 741 (1994)
124. W. Gebhardt, U. Krey: *Phasenübergänge und kritische Phänomene*, 1st ed. (Vieweg, Braunschweig 1980)
125. A. Ghazali, C. Misbah: Phys. Rev. A **46**, 5026 (1992)
126. J.W. Gibbs: T. Conn. Acad. **2**, 382 (1873)
127. M.E. Glicksman: Mater. Sci. Eng. **65**, 45 (1984)
128. M.E. Glicksman, S.P. Marsh: In *Handbook of Crystal Growth*, Vol. **1b**, ed. by D.T.J. Hurle (Elsevier, Amsterdam 1993), p. 1075
129. M.E. Glicksman, M.B. Koss, E.A. Winsa: Phys. Rev. Lett. **73**, 573 (1994)
130. M. E. Glicksman: Microgravity News, NASA **4**, 4 (1997)
131. M. E. Glicksman: private communication (Aachen 2002)
132. R. Glowinski, T.-S. Pan, J. Periaux: Comput. Methods Appl. Mech. Eng. **111**, 283 (1994)
133. G.H. Godunov, V.S. Ryabenkii: *Difference Schemes: An Introduction to the Underlying Theory* (North Holland, Amsterdam 1987)
134. H. Goldstein: *Klassische Mechanik*, 11th ed. (Aula-Verlag, Wiesbaden 1978)
135. G.H. Golub, J.M. Ortega: *Scientific Computing* (Academic Press, Boston 1993)
136. P.M. Gresho, S.T. Chan, M.A. Christon, A.C. Hindmarsh: Int. J. Numer. Methods Fluids **21**, 837 (1995)
137. M. Griebel, T. Dornseifer, T. Neunhofer: *Numerische Simulation in der Strömungsmechanik*, 1st ed. (Vieweg, Braunschweig 1995)

138. B. Grossmann, K.R. Elder, M. Grant, J. M. Kosterlitz: Phys. Rev. Lett. **71**, 3323 (1993)
139. J.D. Gunton, M. San Miguel, P. Sahni: In *Phase Transitions and Critical Phenomena*, Vol.**8**, ed. by C. Domb, J.L. Lebowitz (Academic Press, London 1983), p. 267
140. I. Gyrmati: *Non-Equilibrium Thermodynamics*, 1st ed. (Springer, New York 1970)
141. *Proceedings of the Workshop on Virtual Molecular Beam Epitaxy*, ed. by M.F. Gyure, J.J. Zinck, *Comp. Mater. Sci.* **6**, 113 (1996)
142. R. Haase, H. Schoenert: *Solid–Liquid Equilibrium*, 1st ed. (Pergamon, Oxford 1969)
143. B.I. Halpering, P.C. Hohenberg, S. Ma: Phys. Rev. B **10**, 139 (1974)
144. S.C. Hardy: J. Cryst. Growth **69**, 456 (1984)
145. D.W. Hoffmann, J.W. Cahn: Surf. Sci. **31**, 368 (1972)
146. P.C. Hohenberg, B.I. Halperin: Rev. Mod. Phys. **49**, 435 (1977)
147. G. Horvay, J. W. Cahn: Acta Metall. **9**, 695 (1961)
148. T. Hou, Z. Li, S. Osher, H.-K. Zhao: J. Comput. Phys. **134**, 236 (1997)
149. S.-C. Huang, M. Glicksman: Acta Metall. **29**, 1697 (1981)
150. T. Ihle, H. Müller-Krumbhaar: Phys. Rev. E **49**, 2972 (1994)
151. T. Ihle: Eur. Phys. J. B **16**, 337 (2000)
152. S.E. Ingle, F.H. Horne: J. Chem. Phys. **59**, 5882 (1973)
153. G. P. Ivantsov: Dokl. Akad. Nauk USSR **58**, 1113 (1947)
154. D. Jasnow: Rep. Prog. Phys. **47**, 1061 (1984)
155. J.-H. Jeong, N. Goldenfeld, J. Dantzig: Phys. Rev. E **64**, 041602 (2001)
156. R.A. Johnson, D.M. Belk: AIAA J. **33**, 2305 (1995)
157. D. Juric, G. Tryggvason: J. Comput. Phys. **123**, 127 (1996)
158. L.V. Kale, S. Krishnan: In *Parallel Programming using C++*, ed. by G.V. Wilson, P. Lu (MIT Press, Massachusetts 1996), p. 175
159. N.S. Kalospiros, B.J. Edward, A.N. Beris: Int. J. Heat Mass Trans. **36**, 1191 (1993)
160. N.S. Kalospiros, R. Ocone, G. Astarita, J.H. Meldon: Ind. Eng. Chem. Res. **30**, 851 (1991)
161. S. Kaplun: In *Fluid Mechanis and Singular Perturbations*, ed. by P.A. Lagerstrom, L.N. Howard, C.S. Liu (Academic Press, New York 1967)
162. A. Karma, J.S. Langer: Phys. Rev. A **30**, 3147 (1984)
163. A. Karma, B.G. Kotliar: Phys. Rev. A **31**, 3266 (1985)
164. A. Karma, A. Sarkissian: Phys. Rev. E **47**, 513 (1993)
165. A. Karma: Phys. Rev. E **48**, 3441 (1993)
166. A. Karma: Phys. Rev. E **49**, 2245 (1994)
167. A. Karma, W. Rappel: Phys. Rev. Lett. **77**, 4050 (1996)
168. A. Karma, W.-J. Rappel: Phys. Rev. E **53**, R3017 (1996)
169. A. Karma, W.-J. Rappel: Phys. Rev. E **57**, 4323 (1998)
170. A. Karma, M. Plapp: Phys. Rev. Lett. **81**, 4444 (1998)
171. A. Karma: Phys. Rev. Lett. **87**, 115701 (2001)
172. K. Kassner, C. Misbah, J. Müller, J. Kappe, P. Kohlert: Phys. Rev. E **63**, 036117 (2001)
173. K. Kawasaki, T. Ohta: Prog. Theor. Phys. **67**, 147 (1982)
174. K. Kawasaki, T. Ohta: Prog. Theor. Phys. **68**, 129 (1982)
175. K. Kawasaki, T. Ohta: Physica **118A**, 175 (1983)
176. K. Kawasaki, M.C. Yalabik, J.D. Gunton: Phys. Rev. A **17**, 455 (1978)
177. L.J.T.M. Kemper: J. Chem. Phys. **115**, 6330 (2001)

178. J. Kepler: 'Strena seu de nive sexangula', ed. by G. Tampach (Frankfurt 1611); translated in *A New Year's Gift or on the Six–Cornered Snowflake*, ed. by C. Hardie (Clarendon, Oxford 1966), p. 74
179. D. Kessler, J. Koplik, H. Levine: Phys. Rev. A **30**, 3161 (1984)
180. D. Kessler, H. Levine: Europhys. Lett. **4**, 215 (1987)
181. D. Kessler, J. Koplik, H. Levine: Adv. Phys. **37**, 255 (1988)
182. Y.T. Kim, N. Provatas, N. Goldenfeld, J. Dantzig: Phys. Rev. E **59**, R2546 (1999)
183. Y.T. Kim, N. Goldenfeld, J. Dantzig: Phys. Rev. B **62**, 2471 (2000)
184. M.B. King: *Phase Equilibrium Mixtures*, 1st ed. (Pergamon Press, Oxford 1969)
185. J.S. Kirkaldy, D.J. Young: *Diffusion in the Condensed State* (The Institute of Metals, London 1987)
186. R. Kobayashi: Bull. Jpn. Soc. Ind. Appl. Math. **1**, 22 (1991)
187. R. Kobayashi: Physica D **63**, 410 (1993)
188. R. Kobayashi, J.A. Warren, W.C. Carter: Physica D **140**, 141 (2000)
189. P. Kopczynski, W.-J. Rappel, A. Karma: Phys. Rev. Lett. **77**, 3387 (1996)
190. D.J. Korteweg: Arch. Neerl. Sci. Exactes Nat. Ser. II **6** (1901)
191. M. Kruskal, H. Segur: 'Asymptotics Beyond all Orders in a Model of Dendritic Crystals'. In *Aero. Res. Ass. of Princeton Tech. Memo* (Princeton 1985)
192. R. Kupfermann: *Morphology, Coexistence and Selection in interfacial Pattern Formation*. PhD Thesis, Tel Aviv University, Tel Aviv (1991)
193. J.C. LaCombe, M.B. Koss, M.E. Glicksman: Phys. Rev. Lett. **83**, 2997 (1999)
194. J.C. LaCombe, M.B. Koss, J.E. Frei, C. Giummarra, A.O. Lupulescu, M.E. Glicksman: Phys. Rev. E **65**, 031604 (2002)
195. B. Lafaurie, S. Zaleski, G. Zanetti: J. Comp. Phys. **113**, 134 (1994)
196. P.A. Lagerstrom, D.A. Reinelt: SIAM J. Appl. Math. **44**, 451 (1984)
197. P.A. Lagerstrom: *Matched Asymptotic Expansions*, 1st ed. (Springer, New York 1988)
198. Landoldt-Börnstein, Vol. **IV/5 D** (Springer, New York 1993)
199. J.S. Langer: Rev. Mod. Phys. **52**, 1 (1980)
200. J. Langer: In *Directions in Condensed Matter Physics*, ed. by G. Grinstein, G. Mazenko (World Scientific, Singapore 1986), p. 164
201. J.S. Langer: Phys. Rev. A **36**, 3350 (1987)
202. J. Langer: In *Chance and Matter*, Les Houches Session XLVI, ed. by J. Souletie, J. Vannenimus, R. Stora (North Holland, Amsterdam, 1987), p. 629
203. Y.-W. Lee, R. Ananth, W.N. Gill: Chem. Eng. Comm. **153**, 41 (1996)
204. F.K. LeGoues, M.C. Reuter, J. Tersoff, H. Hammar, R.M. Tromp: Phys. Rev. Lett **73**, 300 (1994)
205. M.-A. Lemieux, G. Kotliar: Phys. Rev. A **36**, 4975 (1987)
206. R.J. LeVeque, Z. Li: SIAM J. Numer. Anal. **31**, No. 4, 1019 (1994)
207. F. Liu, H. Metiu: Phys. Rev. E **49**, 2601 (1994)
208. J. Lowengrub, J. Truskinovsky: Proc. Roy. Soc. Ser. A **454**, 2617 (1998)
209. S. Luckhaus: 'Solidification of alloys and the Gibbs–Thomson law'. Preprint No. 335, Institute of Mathematics, University of Bonn (1993)
210. B.Y. Lyubov: *Kinetic Theory of Phase Transitions* (Metallurgia Publishers, Moskau 1978)
211. A. Madhukar, S. V. Ghaisas: CRC Crit. Rev. Solid State Mater. Sci. **14**, 1 (1988)
212. G.B. McFadden, S.R. Coriell: J. Cryst. Growth **74**, 507 (1986)
213. G.B. McFadden, A.A. Wheeler, R.J. Braun, S.R. Coriell, R.F. Sekerka: Phys. Rev. E **48**, 2016 (1993)
214. G.B. McFadden, A.A. Wheeler, D.M. Anderson: Physica D **144**, 154 (2000)

215. B. Merriman, J. Bence, S. Osher: J. Comput. Phys. **112**, 334 (1994)
216. H. Metui, Y.-T. Lu, Z.Y. Zhang: Science **255**, 1088 (1992)
217. X. Mikic, E.C. Morse: J. Comp. Phys. **61**, 154 (1985)
218. Y.-W. Mo, D.E. Savage, B.S. Schwartzentruber, M.G. Lagally: Phys. Rev. Lett **65** 1020 (1990)
219. P. Morse, H. Feshbach: *Methods of Theoretical Physics*, 1st ed. (McGraw-Hill, New York 1953)
220. J. Müller, M. Grant: Phys. Rev. Lett. **82**, 1736 (1999)
221. H. Müller-Krumbhaar, H. Emmerich, E. Brener, M. Hartmant: Phys. Rev. E **63**, 026304 (2001)
222. W.W. Mullins: In *Metal Surfaces: Structure, Energetics, and Kinetics*, ed. American Society for Metals (1962), p. 17
223. W.W. Mullins, R.F. Sekerka: J. Appl. Physics **35**, 444 (1964)
224. M. Muschol, D. Lui, H. Z. Cummins: Phys. Rev. A **64** (1992) 1038
225. T. Ohta, D. Jasnow, K. Kawasaki: Phys. Rev. Lett. **49**, 1223 (1982)
226. S. Ono, S. Kondo: Encyclopaedia of Physics **10**, Springer (1960)
227. L. Onsager: Phys. Rev. **37**, 405 (1931)
228. S. Osher, J.A. Sethian: J. Comput. Phys. **79**, 12 (1988)
229. N. Palle, J. A. Dantzig: Met. Trans. A **27A**, 707 (1996)
230. R.B. Pember, J.B. Bell, P. Colella: J. Comput. Phys. **120**, 278 (1995)
231. O. Penrose, P.C. Fife: Physica D **43**, 44 (1990)
232. O. Penrose, P.C. Fife: Physica D **69**, 107 (1993)
233. C.S. Peskin: J. Comput. Phys. **25**, 220 (1977)
234. R. Peters, J.S. Langer: Phys. Rev. Lett. **56**, 1948 (1986)
235. O. Pierre–Louis, C. Misbah, Y. Saito, J. Krug, P. Politi: Phys. Rev. Lett. **80**, 4221 (1998)
236. Y. Pomeau, M. Ben Amar: In *Solids far from Equilibrium*, ed. by Godrèche (Cambridge University Press, Cambridge 1991), p. 365
237. I. Prigogine: *Non-Equilibrium Statistical Mechanics*, 2nd ed. (Wiley, New York 1966)
238. N. Provatas, E. Elder, M. Grant: Phys. Rev. B **53**, 6263 (1996)
239. N. Provatas, N. Goldenfeld, J. Dantzig: Phys. Rev. Lett. **80**, 3308 (1998)
240. N. Provatas, N. Goldenfeld, J. Dantzig: J. Comp. Phys. **148**, 265 (1999)
241. J.J. Quirk: ICASE Report No. 92–7, NASA Langley Research Center, Hampton, VA (1992)
242. C. Ratsch, P. Ruggerone, M. Scheffler: In *Surface Diffusion: Atomistic and Collective Processes*, ed. by M.C. Tringides (Plenum, New York 1997), p. 83
243. W.H.S. Rouse: *Great Dialogues of Plato (transl.) – A Mentor Book*, 1st reissue ed. (Signet Classic, New York 1999)
244. J. S. Rowlinson: J. Stat. Phys. **20**, 197 (1979) (an english translation of the original Ph.D. thesis by van der Waals)
245. Y. Saito, G. Goldbeck-Wood, H. Müller–Krumbhaar: Phys. Rev. A **38**, 2148 (1988)
246. A. Schmidt: J. Comp. Phys. **125**, 293 (1996)
247. L.Z. Schlitz, S. Garimella: J. Appl. Phys. **76**, 4863 (1994)
248. D. Schwabe, A. Scharmann: Phys. Bl. **42**, 352 (1986)
249. V. Seetharaman, M.A. Eshelman, R. Trivedi: Acta Metall. **36**, 1175 (1988)
250. R.F. Sekerka, Z. Bi: Physica A **261**, 95 (1998)
251. R.F. Sekerka: In *Proceedings of the Microgravity Materials Science Conference* (Huntsville, Alabama 2000), p. 533
252. R.F. Sekerka, Z. Bi : In *Interfaces for the Twenty-First Century* ed. by M.K. Smith, M.J. Miksis, G.B. McFadden, G.P. Neitzel, D.R. Canright (Imperial College Press, London 2002), p. 147

253. J.A. Sethian: Proc. Nat. Acad. Sci. **93**, 1591 (1996)
254. J.A. Sethian: *Level Set Methods and Fast Marching Methods*, 2nd ed. (Cambridge University Press, Cambridge 1999)
255. M. S. Shephard, P. L. Baehmann, K. R. Grice: Comm. App. Num. Meth. **4**, 379 (1988)
256. W. Shyy: *Computational Modeling for Fluid Flow and Interfacial Transport* (Elsevier, Amsterdam 1994)
257. W. Shyy, H.S. Udaykumar, M.M. Rao, R.W. Smith: *Computational Fluid Dynamics with Moving Boundaries* (Taylor & Francis, Washington, DC, 1996)
258. A.J. Simon, J. Bechhoefer, A. Libchaber: Phys. Rev. Lett. **61**, 2574 (1988)
259. J. Smith: J. Comp. Phys. **39**, 112 (1981)
260. K. Somboonsuk, J.T. Mason, R. Trivedi: Metall. Trans. **15A**, 967 (1984)
261. B.J. Spencer, S.H. Davis, P.W. Voorhees: Phys. Rev. B **47**, 9760 (1993)
262. D.J. Srolovitz: Acta Metallogr. **37**, 621 (1988)
263. H.E. Stanley: *Phase Transitions and Critical Phenomena*, 1st ed. (Oxford Univ. Press, New York 1971)
264. J.L. Steger: In *Proceedings of the Computational Fluid Dynamics Symposium on Aeropropulsion, NASA CP–3078* (Natl. Aeronautics & Space Admin., Washington, DC, 1991), p. 1
265. Z. Suo: Advances in Appl. Mechs. **33**, 193 (1997)
266. M. Sussman, P. Smereka, S. Osher: J. Comput. Phys. **114**, 146 (1994)
267. J.E. Taylor: Acta Met. **40**, 1475 (1992)
268. J.E. Taylor, J.W. Cahn: Acta Met. **42**, 1045 (1994)
269. J.E. Taylor, J.W. Cahn: J. Stat. Phys. **77**, 183 (1994)
270. D. E. Temkin: Dokl. Akad. Nauk SSSR **132**, 1307 (1960)
271. D. ter Haar: *Collected Papers of Landau* (Gordon and Breach, Washington, DC, 1967)
272. R. Tönhardt, G. Amberg: J. Cryst. Growth **194**, 406 (1998)
273. J.C. Tolendano, P. Tolendano: *The Landau Theory of Phase Transitions* (World Scientific, Singapore 1987)
274. R. Trivedi: Metall. Trans. **15A**, 977 (1984)
275. R. Trivedi, J.A. Sekhan, V. Seetharaman: Metall. Trans. **A20**, 769 (1989)
276. J.Y. Tsao: *Materials Fundamentals of Molecular Beam Epitaxy* (Academic Press, Boston 1993)
277. H. S. Udaykumar, W. Shyy: Numer. Heat Transfer B **27**, 127 (1995)
278. H.S. Udaykumar, W. Shyy: Int. J. Heat Mass Transfer **38**, 2057 (1995)
279. H.S. Udaykumar, W. Shyy, M M. Rao: Int. J. Numer. Methods Fluids **22**, 691 (1996)
280. H.S. Udaykumar, H.-C. Kan, W. Shyy, R. Tran-Son-Tay: J. Comput. Phys. **137**, 366 (1997)
281. H.S. Udaykumar, R. Mittal, W. Shyy: J. Comput. Phys. **153**, 535 (1999)
282. A. Umantsev, A.L. Roitburd: Sov. Phys. **30** 651 (1988)
283. A.P. Umantsev: J. Chem. Phys. **96**, 606 (1992)
284. A. Umantsev, S.H. Davis: Phys. Rev. A **45** 7195 (1992)
285. S.O. Unverdi, G. Tryggvason: J. Comput. Phys. **100**, 25 (1992)
286. V. Venkatakrshnan: AIAA J. **34**, 533 (1996)
287. D.D. Vvendensky, A. Zangwill, C.N. Luse, M.R. Wilby: Phys. Rev. E **48**, 852 (1993)
288. J. Villain: J. Phys. I **1**, 19 (1991)
289. S.-L. Wang, R.F. Sekerka, A.A. Wheeler, B.T. Murray, S.R. Coriell, R.B. Braun, G.B. McFadden: Physica D **69**, 189 (1993)
290. S.-L. Wang, R.F. Sekerka: Phys. Rev. E **53**, 3760 (1996)
291. J.A. Warren, W.J. Boettinger: Acta Metall. Mater. A **43**, 689 (1995)

292. J.A. Warren, B.T. Murray: Mod. Simul. Mater. Sci. Eng. **4**, 215 (1996)
293. J.D. Weeks, G.H. Gilmer: Adv. Chem. Phys. **40**, 157 (1979)
294. H. Wenzel: In *Physikalische Grundlagen metallischer Werkstoffe*, 13th IFF-Spring School at Research Center Juelich (1982)
295. A.A. Wheeler, W.J. Boettinger, G.B. McFadden: Phys. Rev. A **45**, 7424 (1992)
296. A.A. Wheeler, W.J. Boettinger, G.B. McFadden: Phys. Rev. E **47**, 1893 (1993)
297. A.A. Wheeler, B.T. Murray, R.J. Schaefer: Physica D **66**, 243 (1993)
298. A.A. Wheeler, G.B. McFadden, W.J. Boettinger: Proc. Royal Soc. London A **452**, 495 (1996)
299. D.J. Wollkind, R.N. Maurer: J. Cryst. Growth **42**, 24 (1977)
300. G. Wulff: Z. Kristallogr. Mineral **34**, 449 (1901)
301. L. Zhang, S. Garimella: J. Appl. Phys. **74**, 2494 (1993)
302. H.K. Zhao, T. Chan, B. Merriman, S. Osher: J. Comput. Phys. **127**, 179 (1996)
303. O. C. Zienkiewicz, J. Z. Zhu: Int. J. Numer. Meth. Eng. **24**, 337 (1987)

Index

G-list   156
H^{-1}–norm   48
L^2–norm   48

triangular elements   156

ADI   121, 133
adiabatic approximation   133, 149
advective term   102
affinities   33
Allen–Cahn equation   5, 40
alloy solidification   142, 143
anisotropic surface tension   9, 47, 104, 106
anti-phase domain coarsening   40
asymptotic analysis   4, 59
asymptotic matching method   59
attachment kinetics   73, 102

balance law formulations   49
binary alloy   55, 147
binary mixture   21
Boltzmann constant   75
boundary-value problem   60
bulk energies   55
bulk modulus   132
bulk region   2
buoyancy   120, 123

Cahn–Hilliard equation   5, 39, 54
– generalized anisotropic   48
– viscous   40
Cahn–Hoffmann capillary vector   47, 51
capillary action   3
capillary length   3, 37, 75, 108, 147
capillary tensor   98
chemical potential   21, 72, 84, 149
chemical rate constants   31
child element   155
coalescence   143
coarsening   135

coexistence regime   24, 79
computation   143
– computational efficiency   154
– computational techniques   154
– computational universality   151
confluent hypergeometric functions   114
conodal line   55
conservation of mass   97
conserved variable approach   28
constant mobility   39
constants of integration   64
constitutive relations   31, 52
continuity equation   33
continuity of tangential components   109
continuity of temperature   106, 110
continuum description   72
continuum equations   69
continuum field models   2
convection limited growth   99
correlation length   2
CPU time   151, 160, 163
cross coupling effects   37
crystalline anisotropy   71
crystalline films   69
crystallization   143
crystallization temperature   9, 106
curvature   9, 47, 67, 78, 102, 106, 110
curvilinear coordinates   84
cusp like solution   135
cylindric coordinates   111

dendritic growth   9, 78, 97
dendritic selection theory   104
dendritic solidification   3
dendritic tip velocity   142
density change flow   103, 104
depending variables   19
deposition time   140
desorptionless epitaxial growth   136
diffuse interface   1, 8

diffuse interface model   39, 59, 78
– for hydrodynamically influenced
  dendritic growth   98
diffusion constant   9
diffusion limited dendritic growth
  104, 142
diffusion limited growth   46, 99, 142
diffusion–relaxation process   71
directional solidification   10
displacement field   134
dissipative boundary force   120
dividing surface   7
double–tangent construction   27
double–well potential   41, 74, 100, 152
doublon morphology   10
doublon–dendrite transition   10
Dufour effect   36

Ehrlich–Schwoebel effect   71
elastic coefficients   31
elastic energy   132
elastic interaction   69
electrodeposition   99, 142
energy
– conservation   110
– density   28, 74, 100, 147
– flux   73
entropic confinement   70
entropy   32
– density   45, 74, 100
– flux   73
– functional   45, 68, 143
– production   49, 74, 99
epitaxial growth   5, 68, 99, 131
equilibrium thermodynamics   19, 31
error estimator function   159
Euler scheme   141
Euler–Lagrange equation   34, 53
Eulerian–Lagrangian methods   14
eutectic alloy solidification   78, 143
extended domains   62
extension theorem   61
extensive quantities
– conserved   33
– non-conserved   35
extremal principle   21, 31

Fast Marching Method   16
finite surface   2
finite-size effects   123
first principle considerations   39
first-principle calculations   72
fluctuations, thermal   122

free energy density   39, 132
free energy functional   2, 78
free-boundary problem   152
– two-sided   75

Galerkin finite element method   155
Galilean transformation   98
generalized analysis   4, 59, 78, 81
Gibbs free energy   19, 21, 32, 99, 148
Gibbs surface   84
Gibbs–Duhem relation   9
Gibbs–Thomson relation   9, 76, 91, 97,
  102, 106, 110, 143, 146
governing variables   19
gradient expansion   33
gradient flow method   47
grain boundaries   144
grain growth   55
gravity   140
Greens function   81, 83, 115
growth rate   38

Helmholtz free energy   19, 44
Hermite polynomials   114
high performance computing   141
homogeneous heat conduction   113
homogenization   144
Hooke's law   132

image processing   142
implementation   145
inhomogeneous heat conduction   115
inhomogeneous states   42
initial value problem   63
inner energy density   46
inner energy functional   68, 73
inner entropy functional   73
inner product   47
inner solution   67, 87
instantaneous relaxation   39
interfacial Peclet number   81, 152
irregular perturbation analysis   104
irregular perturbation expansion   118
irreversible thermodynamics   2, 31, 49,
  140
island growth   69
island morphology   134
isothermal asymptotics   3

kinematic viscosity   106
kinetic attachment   71
kinetic coefficient   147, 149
Kinetic Monte Carlo   69
kinetic principles   31

Korteweg stresses   98
Kummer's equation   113, 117

Lagrangian parameter   33
Landau–Ginzburg–Wilson functional
    42
Landau–Khalatnikov ansatz   39
latent heat   9, 100, 106
lattice model   42
layer-by-layer growth   69
level set method   15
limit of absolute stability   12
liquid crystals   4
liquidus line   12, 23, 136
LPE   135
Lyapunov function   31

MAC   122
Markovian system   33
mass balance   72, 76, 105, 109
masslumping   160
matching   59, 63, 118
MBE   135
mean-field approximation   43
measurable quantities   153
melting temperature   100
mesh refinement   155
metastable phase   55, 81
method of matched asymptotic
    expansion   59, 67, 142
microgravity experiment   105, 120, 142
microscopic derivation   43
microsegregation   143
midpoint scheme   133
miscibility gap   12, 87
misfit strain   132
mobility coefficient   74
Model C   44
model systems   21
moving boundary problems   12, 31
Mullins surface diffusion equation   48
multi-grid algorithm   145
multi-scale approach   144
multicomponent material   99
multicomponent–multiphase system
    23

natural variables   19
Navier–Stokes equations   72, 97
Newtonian liquid   98
no-slip condition   97, 102, 106, 120, 122
non-classical flux   74
non-conserved variable approach   28
nucleation   1

numerical issues   5, 145
numerically tractable models   131

Onsager coefficients   33
Onsager reciprocity relations   33
Onsager theory   31
order parameter   1, 39, 99, 144, 152
order–disorder phase boundaries   144
order/disorder transitions   78
organic materials   127
Ostwald ripening   78, 143
outer solution   67, 87
overlap domains   64

parabolic coordinates   108
parabolic cylinder functions   114
paradigmatic problem   142
parallelization   141, 154
parasite current   122
parent element   155
Peclet number   108
per unit mass formula   100
per unit volume formula   100
periodic perturbation   37
perturbative analysis   126
phase diagram   19, 79
phase separation   55
phase transition phenomena   1
phase transitions   39, 143
phase variable   40
phase-field   78
– approach   132
– equation   78
– model   2, 28
– parameters   148
– variable   32, 38
phenomenological theory   5, 31, 45
Pochhammer symbol   114
Poisson equation   48, 162
Poisson ratio   132
polymer systems   142
Prandtl matching condition   62
Prandtl number   101, 107
preconditioning   163
predictor-corrector method   161
principle of minimal energy dissipation
    33
projection operator   78, 115

quadtree element data structure   155
quasi-incompressible system   99
quench   79

rectangular elements   156

reduced gravity   123
reduced phase diagram   24
regions of uniform validity   60
regular perturbation analysis   104
regular perturbation expansion   112
Reynolds transport theorem   50
rippled morphology   134
roughening   69

scalable algorithm   155
SCN   104
selected tip velocity   118
selection problem of dendritic growth
  97
separation of variables   113
shape function   160, 162
sharp interface   1, 8
sharp interface approach   131
sharp interface limit   78, 89
sharp interface models   59
SIAPM   154, 161
side branching effects   143
silicon   104
smooth tip condition   113
solid–liquid interface   97
solidification   47, 99
solidus line   12, 23, 136
solute   99, 136
solute trapping   143
solvability condition   77, 89, 118, 150
solvability theory   118
Soret effect   36
specific heat   9, 105, 146
spinodal decomposition   1, 40, 54, 78
spinodal line   55
state variables   19
steady state solutions   110
Stefan condition   9, 92, 97, 102, 146
Stefan problem   68
step flow growth   69
strain energy relaxation   134
strain tensor   132
strained surfaces   131
sub-element   155
subsidiary conditions   19
superconducting materials   142

supercooling   146
surface energy   55, 148
surface excess   7
surface quantity   1
surface tension   111, 118

terrace–step–kink model   7
thermocapillary convection   124
thermodynamic consistency   2, 28, 31,
  45, 47, 73, 78, 145
thermodynamic equilibrium   19, 31,
  33, 53
thermodynamic functional   31
thermodynamic potential   19, 28, 33,
  38, 53
thermodynamic state   19
thermodynamic variables   53
thin film growth   99
thin interface analysis   73, 81
thin interface limit   3, 89, 143
tip radius   107
Tolman–Buff effect   148
transparent materials   10
triple junction   57
two-sided epitaxial growth problem
  131

undercooled melt   3, 146
universality   43, 46, 145, 151

validation   2, 75, 141
variational principles   2, 49
variational problem   53
velocity selection   103
virtual displacement   34
viscous stress tensor   52
vorticity
– equation   107
– scalar   107

weakly contaminated melt   24
Weber's equation   114
WKB formalism   118
Wronskian   116

Young's modulus   132

Lecture Notes in Physics

For information about Vols. 1–567
please contact your bookseller or Springer-Verlag

Vol. 568: P. A. Vermeer, S. Diebels, W. Ehlers, H. J. Herrmann, S. Luding, E. Ramm (Eds.), Continuous and Discontinuous Modelling of Cohesive-Frictional Materials. XIV, 307 pages. 2001.

Vol. 569: M. Ziese, M. J. Thornton (Eds.), Spin Electronics. XVII, 493 pages. 2001.

Vol. 570: S. G. Karshenboim, F. S. Pavone, F. Bassani, M. Inguscio, T. W. Hänsch (Eds.), The Hydrogen Atom: Precision Physics of Simple Atomic Systems. XXIII, 293 pages. 2001.

Vol. 571: C. F. Barenghi, R. J. Donnelly, W. F. Vinen (Eds.), Quantized Vortex Dynamics and Superfluid Turbulence. XXII, 455 pages. 2001.

Vol. 572: H. Latal, W. Schweiger (Eds.), Methods of Quantization. XI, 224 pages. 2001.

Vol. 573: H. M. J. Boffin, D. Steeghs, J. Cuypers (Eds.), Astrotomography. XX, 434 pages. 2001.

Vol. 574: J. Bricmont, D. Dürr, M. C. Galavotti, G. Ghirardi, F. Petruccione, N. Zanghi (Eds.), Chance in Physics. XI, 288 pages. 2001.

Vol. 575: M. Orszag, J. C. Retamal (Eds.), Modern Challenges in Quantum Optics. XXIII, 405 pages. 2001.

Vol. 576: M. Lemoine, G. Sigl (Eds.), Physics and Astrophysics of Ultra-High-Energy Cosmic Rays. X, 327 pages. 2001.

Vol. 577: I. P. Williams, N. Thomas (Eds.), Solar and Extra-Solar Planetary Systems. XVIII, 255 pages. 2001.

Vol. 578: D. Blaschke, N. K. Glendenning, A. Sedrakian (Eds.), Physics of Neutron Star Interiors. XI, 509 pages. 2001.

Vol. 579: R. Haug, H. Schoeller (Eds.), Interacting Electrons in Nanostructures. X, 227 pages. 2001.

Vol. 580: K. Baberschke, M. Donath, W. Nolting (Eds.), Band-Ferromagnetism: Ground-State and Finite-Temperature Phenomena. IX, 394 pages. 2001.

Vol.581: J. M. Arias, M. Lozano (Eds.), An Advanced Course in Modern Nuclear Physics. XI, 346 pages. 2001.

Vol.582: N. J. Balmforth, A. Provenzale (Eds.), Geomorphological Fluid Mechanics. X, 579 pages. 2001.

Vol.583: W. Plessas, L. Mathelitsch (Eds.), Lectures on Quark Matter, XIII, 334 pages. 2002.

Vol.584: W. Köhler, S. Wiegand (Eds.), Thermal Nonequilibrium Phenomena in Fluid Mixtures. XVII, 470 pages. 2002.

Vol.585: M. Lässig, A. Valleriani (Eds.), Biological Evolution and Statistical Physics. XI, 337 pages. 2002.

Vol.586: Y. Auregan, A. Maurel, V. Pagneux, J.-F. Pinton (Eds.), Sound–Flow Interactions. XVI, 286 pages. 2002

Vol.587: D. Heiss (Ed.), Fundamentals of Quantum Information. Quantum Computation, Communication, Decoherence and All That. XIII, 265 pages. 2002.

Vol.588: Y. Watanabe, S. Heun, G. Salviati, N. Yamamoto (Eds.), Nanoscale Spectroscopy and Its Applications to Semiconductor Research. XV, 306 pages. 2002.

Vol.589: A. W. Guthmann, M. Georganopoulos, A. Marcowith, K. Manolakou (Eds.), Relativistic Flows in Astrophysics. XII, 241 pages. 2002

Vol.590: D. Benest, C. Froeschlé (Eds.), Singularities in Gravitational Systems. Applications to Chaotic Transport in the Solar System. XI, 215 pages. 2002

Vol.591: M. Beyer (Ed.), CP Violation in Particle, Nuclear and Astrophysics. XI, 334 pages. 2002

Vol.592: S. Cotsakis, L. Papantonopoulos (Eds.), Cosmological Crossroads. An Advanced Course in Mathematical, Physical and String Cosmology. XVI, 477 pages. 2002

Vol.593: D. Shi, B. Aktaş, L. Pust, F. Mikhailov (Eds.), Nanostructured Magnetic Materials and Their Applications. XII, 289 pages. 2002

Vol.594: S. Odenbach (Ed.),Ferrofluids. Magnetical Controllable Fluids and Their Applications. XI, 253 pages. 2002

Vol.595: C. Berthier, L. P. Lévy, G. Martinez (Eds.), High Magnetic Fields. Applications in Condensed Matter Physics and Spectroscopy. X, 493 pages. 2002

Vol.596: F. Scheck, H. Upmeier, W. Werner (Eds.), Noncommutative Geometry and the Standard Model of Elememtary Particle Physics. XII, 346 pages. 2002

Vol.597: P. Garbaczewski, R. Olkiewicz (Eds.), Dynamics of Dissipation. X, 512 pages. 2002

Vol.598: K. Weiler (Ed.), Supernovae and Gamma-Ray Bursters. X, 350 pages. 2002

Vol.600: K. Mecke, D. Stoyan (Eds.), Morphology of Condensed Matter. Physcis and Geometry of Spatial Complex Systems. XVIII, 439 pages. 2002

Vol.601: F. Mezei, C. Pappas, T. Gutberlet (Eds.), Neutron Spin Echo Spectroscopy. Basics, Trends and Applications. XV, 345 pages. 2003

Vol.602: T. Dauxois, S. Ruffo, E. Arimondo (Eds.), Dynamics and Thermodynamics of Systems with Long Range Interactions. XI, 492 pages. 2002

Vol.603: C. Noce, A. Vecchione, M. Cuoco, A. Romano (Eds.), Ruthenate and Rutheno-Cuprate Materials. Superconductivity, Magnetism and Quantum Phase. XXIII, 329 pages. 2002

Vol.604: J. Frauendiener, H. Friedrich (Eds.), The Conformal Structure of Space-Time: Geometry, Analysis, Numerics. XIV 373pages. 2002

Vol.605: G. Ciccotti, M. Mareschal, P. Nielaba (Eds.), Bridging Time Scales: Molecular Simulations for the Next Decade. XXVI, 500 pages. 2002

Vol.608: F. Courbin, D. Minniti (Eds.), Gravitational Lensing:An Astrophysical Tool. XI, 191 pages. 2002

Vol.611: A. Buchleitner, K. Hornberger (Eds.), Coherent Evolution in Noisy Environments. XV, 293 pages. 2002

Vol.613 K. Porsezian, V.C. Kuriakose, (Eds.), Optical Solitons. Theoretical and Experimental Challenges. XI, 406 pages. 2003

Monographs

For information about Vols. 1–30
please contact your bookseller or Springer-Verlag

Vol. m 31 (Corr. Second Printing): P. Busch, M. Grabowski, P.J. Lahti, Operational Quantum Physics. XII, 230 pages. 1997.

Vol. m 32: L. de Broglie, Diverses questions de mécanique et de thermodynamique classiques et relativistes. XII, 198 pages. 1995.

Vol. m 33: R. Alkofer, H. Reinhardt, Chiral Quark Dynamics. VIII, 115 pages. 1995.

Vol. m 34: R. Jost, Das Märchen vom Elfenbeinernen Turm. VIII, 286 pages. 1995.

Vol. m 35: E. Elizalde, Ten Physical Applications of Spectral Zeta Functions. XIV, 224 pages. 1995.

Vol. m 36: G. Dunne, Self-Dual Chern-Simons Theories. X, 217 pages. 1995.

Vol. m 37: S. Childress, A.D. Gilbert, Stretch, Twist, Fold: The Fast Dynamo. XI, 406 pages. 1995.

Vol. m 38: J. González, M. A. Martín-Delgado, G. Sierra, A. H. Vozmediano, Quantum Electron Liquids and High-Tc Superconductivity. X, 299 pages. 1995.

Vol. m 39: L. Pittner, Algebraic Foundations of Non-Com-mutative Differential Geometry and Quantum Groups. XII, 469 pages. 1996.

Vol. m 40: H.-J. Borchers, Translation Group and Particle Representations in Quantum Field Theory. VII, 131 pages. 1996.

Vol. m 41: B. K. Chakrabarti, A. Dutta, P. Sen, Quantum Ising Phases and Transitions in Transverse Ising Models. X, 204 pages. 1996.

Vol. m 42: P. Bouwknegt, J. McCarthy, K. Pilch, The W3 Algebra. Modules, Semi-infinite Cohomology and BV Algebras. XI, 204 pages. 1996.

Vol. m 43: M. Schottenloher, A Mathematical Introduction to Conformal Field Theory. VIII, 142 pages. 1997.

Vol. m 44: A. Bach, Indistinguishable Classical Particles. VIII, 157 pages. 1997.

Vol. m 45: M. Ferrari, V. T. Granik, A. Imam, J. C. Nadeau (Eds.), Advances in Doublet Mechanics. XVI, 214 pages. 1997.

Vol. m 46: M. Camenzind, Les noyaux actifs de galaxies. XVIII, 218 pages. 1997.

Vol. m 47: L. M. Zubov, Nonlinear Theory of Dislocations and Disclinations in Elastic Body. VI, 205 pages. 1997.

Vol. m 48: P. Kopietz, Bosonization of Interacting Fermions in Arbitrary Dimensions. XII, 259 pages. 1997.

Vol. m 49: M. Zak, J. B. Zbilut, R. E. Meyers, From Instability to Intelligence. Complexity and Predictability in Nonlinear Dynamics. XIV, 552 pages. 1997.

Vol. m 50: J. Ambjørn, M. Carfora, A. Marzuoli, The Geometry of Dynamical Triangulations. VI, 197 pages. 1997.

Vol. m 51: G. Landi, An Introduction to Noncommutative Spaces and Their Geometries. XI, 200 pages. 1997.

Vol. m 52: M. Hénon, Generating Families in the Restricted Three-Body Problem. XI, 278 pages. 1997.

Vol. m 53: M. Gad-el-Hak, A. Pollard, J.-P. Bonnet (Eds.), Flow Control. Fundamentals and Practices. XII, 527 pages. 1998.

Vol. m 54: Y. Suzuki, K. Varga, Stochastic Variational Approach to Quantum-Mechanical Few-Body Problems. XIV, 324 pages. 1998.

Vol. m 55: F. Busse, S. C. Müller, Evolution of Spontaneous Structures in Dissipative Continuous Systems. X, 559 pages. 1998.

Vol. m 56: R. Haussmann, Self-consistent Quantum Field Theory and Bosonization for Strongly Correlated Electron Systems. VIII, 173 pages. 1999.

Vol. m 57: G. Cicogna, G. Gaeta, Symmetry and Perturbation Theory in Nonlinear Dynamics. XI, 208 pages. 1999.

Vol. m 58: J. Daillant, A. Gibaud (Eds.), X-Ray and Neutron Reflectivity: Principles and Applications. XVIII, 331 pages. 1999.

Vol. m 59: M. Kriele, Spacetime. Foundations of General Relativity and Differential Geometry. XV, 432 pages. 1999.

Vol. m 60: J. T. Londergan, J. P. Carini, D. P. Murdock, Binding and Scattering in Two-Dimensional Systems. Applications to Quantum Wires, Waveguides and Photonic Crystals. X, 222 pages. 1999.

Vol. m 61: V. Perlick, Ray Optics, Fermat's Principle, and Applications to General Relativity. X, 220 pages. 2000.

Vol. m 62: J. Berger, J. Rubinstein, Connectivity and Superconductivity. XI, 246 pages. 2000.

Vol. m 63: R. J. Szabo, Ray Optics, Equivariant Cohomology and Localization of Path Integrals. XII, 315 pages. 2000.

Vol. m 64: I. G. Avramidi, Heat Kernel and Quantum Gravity. X, 143 pages. 2000.

Vol. m 65: M. Hénon, Generating Families in the Restricted Three-Body Problem. Quantitative Study of Bifurcations. XII, 301 pages. 2001.

Vol. m 66: F. Calogero, Classical Many-Body Problems Amenable to Exact Treatments. XIX, 749 pages. 2001.

Vol. m 67: A. S. Holevo, Statistical Structure of Quantum Theory. IX, 159 pages. 2001.

Vol. m 68: N. Polonsky, Supersymmetry: Structure and Phenomena. Extensions of the Standard Model. XV, 169 pages. 2001.

Vol. m 69: W. Staude, Laser-Strophometry. High-Resolution Techniques for Velocity Gradient Measurements in Fluid Flows. XV, 178 pages. 2001.

Vol. m 70: P. T. Chruściel, J. Jezierski, J. Kijowski, Hamiltonian Field Theory in the Radiating Regime. VI, 172 pages. 2002.

Vol. m 71: S. Odenbach, Magnetoviscous Effects in Ferrofluids. X, 151 pages. 2002.

Vol. m 72: J. G. Muga, R. Sala Mayato, I. L. Egusquiza (Eds.), Time in Quantum Mechanics. XII, 419 pages. 2002.

Vol. m 73: H. Emmerich, The Diffuse Interface Approach in Materials Science. VIII, 178 pages. 2003

Druck: betz-druck GmbH, D-64291 Darmstadt
Verarbeitung: Buchbinderei Schäffer, D-67269 Grünstadt